Werner Sobek
Light Works

avedition

Frank Heinlein
Maren Sostmann

Werner Sobek
Light Works

Structures

Wherever the structure of a building becomes visible, Werner Sobek aims to integrate this structure perfectly within the overall architectural intention. This requires the engineer to have a deep understanding of the architectural intention – and the architect to have an understanding of the possibilities and limitations of the supporting structure. It requires cross-disciplinary skills and interdisciplinary ability.

Achieving harmony between function, materials, form, and design of all building components is one of the greatest goals of architecture. This is possible only if the boundaries between architect and engineer are dissolved. Werner Sobek develops structures built according to constructive logic, design quality, lightness, and transparency. But the studio not only develops such structures, it also adapts them to fit harmoniously into the overall context. For Werner Sobek, the designing of a structure involves a reduction to the quintessential, at all times testing the limits of what is physically and technically feasible. Minimized structures with very wide spans and an extremely high degree of design quality are the result of this approach.

Überall dort, wo die tragenden Teile eines Gebäudes hervortreten, wo die Konstruktion eines Gebäudes sichtbar ist, kommt es auf die perfekte Einpassung des Tragwerks in den Gesamtkontext, auf die Abstimmung von Konstruktion und architektonischer Intention an. Dies erfordert ein tief greifendes Verständnis des Ingenieurs für die architektonische Intention und ein Verständnis des Architekten für Möglichkeiten und Grenzen der tragenden Konstruktion. Es erfordert eine die Disziplinen übergreifende Qualifikation und interdisziplinäres Können.

Die Übereinstimmung zwischen Funktion, Materialität, Form und konstruktiver Ausbildung von Bauteil und Bauwerk ist eines der großen Ziele der Architektur. Möglich wird sie nur durch die Auflösung der Grenzen zwischen Architekt und Ingenieur. Werner Sobek entwickelt gebaute Strukturen nach konstruktiver Logik, Gestaltqualität, Leichtigkeit und Transparenz und passt diese harmonisch in den Gesamtkontext ein. Das Entwerfen einer Tragstruktur bedeutet hier eine Reduktion auf das Notwendigste, ein Arbeiten an den Grenzen des physikalisch und technisch Machbaren. Minimierte Strukturen mit großen Spannweiten und höchster Gestaltqualität sind das Ergebnis dieses Ansatzes.

Post Tower
Architect: Murphy/Jahn, Chicago, USA
Planning time: 1999–2002
Construction time: 2000–2003
Tasks completed: full engineering service except site observation for structural engineering of load-bearing structure, special structures, and facades
Client: Deutsche Post AG, Bonn, Germany

Multi-story buildings are the landmarks of modern cities. In a very obvious way, but also in a figurative sense, they embody the human aspiration to something higher. They symbolize the prosperity and the importance of a city or a company, thus exercising a formative influence on the public perception of the user of the building. Multi-story buildings are, therefore, of essential importance in modern architecture.

Multi-story buildings present a special challenge not only for architects. They are also (perhaps even to a greater degree) a challenge to the engineer in charge of the structure, the facade or the energy strategy for such a building. The wind loads at a height of several hundred meters, the risk of earthquakes, the minimization of the materials used combined with a maximization of the transparency of facade and structure – all this must be considered. The planning of a multi-story building is an extremely complex process, which demands extensive and detailed specialized knowledge of the planners.

Werner Sobek has designed and planned structures and facades for numerous multi-story buildings in Europe, America, and Asia. All of these buildings are characterized by their lightness and transparency. A sturdy yet scarcely perceptible structure is just as much a characteristic trademark of these buildings as their virtually dematerialized facades. These facades are not only crucial for the building design – they are also an important part of the climate and energy management.

Hochhäuser sind die Wahrzeichen moderner Städte. Sie verkörpern – ganz augenscheinlich, aber auch im übertragenen Sinne – das menschliche Streben nach Höherem. Sie symbolisieren den Wohlstand und die Bedeutung einer Stadt oder eines Unternehmens. Hochhäuser sind deshalb von essenzieller Wichtigkeit in der modernen Architektur.

Nicht nur für Architekten sind Hochhäuser eine besondere Herausforderung. Sie sind auch – vielleicht sogar in noch größerem Maße – eine Herausforderung für den Ingenieur, der das Tragwerk, die Fassade oder das Energiekonzept eines solchen Gebäudes plant. Die Windlasten in mehreren hundert Metern Höhe, das Risiko von Erdbeben, die Minimierung der verwendeten Materialien bei gleichzeitiger Maximierung der Transparenz von Fassade und Tragwerk, all das muss berücksichtigt werden. Die Planung eines Hochhauses ist ein extrem komplexer Prozess, der den beteiligten Planern ein sehr umfangreiches und detailliertes Fachwissen abverlangt.

Werner Sobek hat Tragwerke und Fassaden für zahlreiche Hochhäuser in Europa, Amerika und Asien konzipiert und geplant. Alle diese Hochhäuser zeichnen sich durch ihre Leichtigkeit und Transparenz aus. Ein robustes, aber dennoch kaum wahrnehmbares Tragwerk gehört eben-so zu ihren Markenzeichen wie die nahezu entmaterialisierten Fassaden, die ein wichtiger Teil des Klima- und Energiemanagements sind.

Post Tower

The first multi-story building where the synthesis of high transparency, perfect functionality and excellent component design could be fully realized was the new head office of Deutsche Post in Bonn. This building, which is 162 m high and consists of two segments of a circle offset against each other, is surrounded by a filigree two-shell facade. Thanks to this facade, it is possible to ventilate and cool the building naturally. In addition, the users can adjust the supply of fresh air individually in all offices and open the windows. Other remarkable features include the projecting wing-walls, which are stabilized by filigree wind pins, and the penthouse area which is surrounded by a glass panel 11 m high.

Das erste Hochhaus, bei dem die Synthese aus hoher Transparenz, perfekter Funktionalität und exzellenter Bauteilgestaltung voll zum Tragen kommen konnte, war die neue Hauptverwaltung der Deutschen Post in Bonn. Dieses 162 m hohe, aus zwei gegeneinander versetzten Kreisabschnitten bestehende Gebäude ist von einer filigranen zweischaligen Fassade umgeben. Dank dieser Fassade ist es möglich, das Gebäude natürlich zu belüften und zu kühlen. Darüber hinaus können die Nutzer in allen Büros die Frischluftzufuhr individuell regulieren und die Fenster öffnen. Bemerkenswert sind auch die weit auskragenden Wingwalls, die von filigranen Windnadeln stabilisiert werden, sowie der Penthouse-Bereich, der von einer 11 m hohen Glaswand umgeben ist.

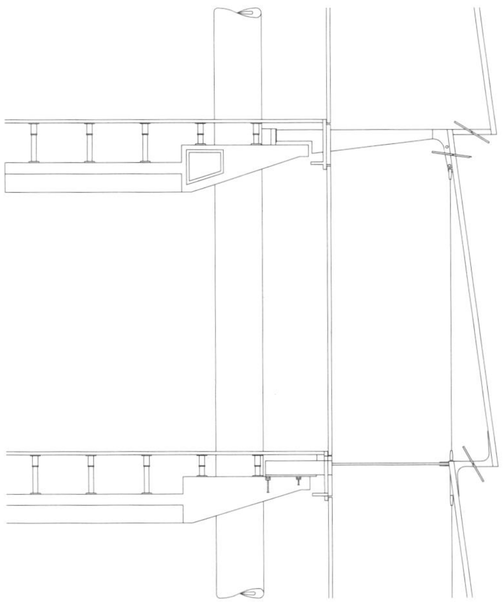

Post Tower

HighLight Towers
Architect: Murphy/Jahn, Chicago, USA
Planning time: 2001–2002
Construction time: 2002–2004
Tasks completed: full engineering service except site observation for structural engineering of load-bearing structure, special structures, and facades
Client: KanAm, Munich, Germany

The HighLight Towers in Munich are a further example of an extreme reduction of structure. This consistently lightweight construction not only minimizes the mass of the building materials employed, but also results in a high level of transparency and openness of the floor plans. The ensemble consists of two offset multi-story building slices, with heights of 126 and 109 m respectively. Both buildings are approx. 80 m long and 14 m wide. The relationship of height to width is therefore almost 1:10 – a value which has rarely been achieved in multi-story buildings. Since the two multi-story building slices are developed as statically-isolated systems, the connecting bridges between the buildings can be moved or removed at will, in order to adapt to any changes in user requirements.

Die HighLight Towers in München sind ein weiteres Beispiel für eine extreme Reduzierung des Tragwerks. Dieser konsequente Leichtbau minimiert nicht nur die Masse der eingesetzten Baumaterialien, sondern führt auch zu einer großen Transparenz und Offenheit der Grundrisse. Das Ensemble besteht aus zwei versetzten Hochhausscheiben mit einer Höhe von 126 bzw. 109 m. Beide Gebäude sind je ca. 80 m lang und 14 m breit. Das Verhältnis von Höhe zu Breite beträgt also nahezu 1:10 – ein bei Hochhäusern bislang kaum erreichter Wert. Da die beiden Hochhausscheiben als statisch getrennte Systeme ausgebildet sind, können die Verbindungsbrücken zwischen den Gebäuden beliebig versetzt oder entfernt werden, um sich möglicherweise ändernden Nutzerbedürfnissen anzupassen.

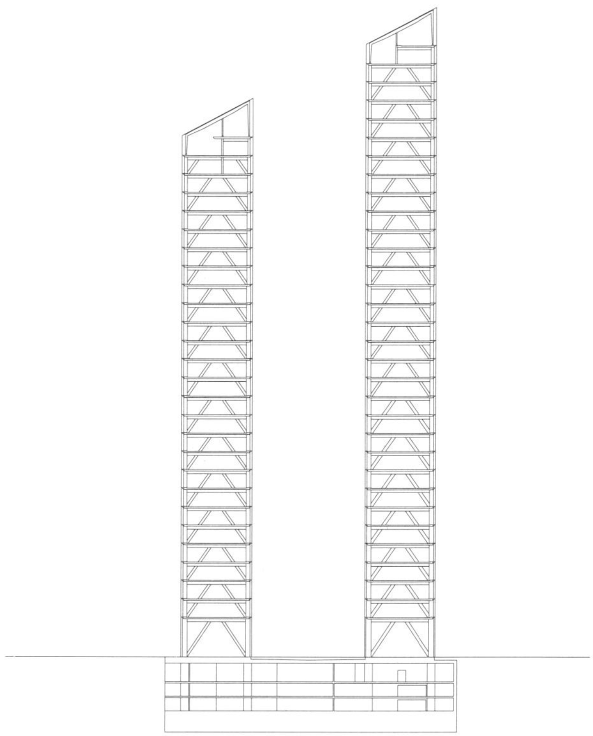

HighLight Towers

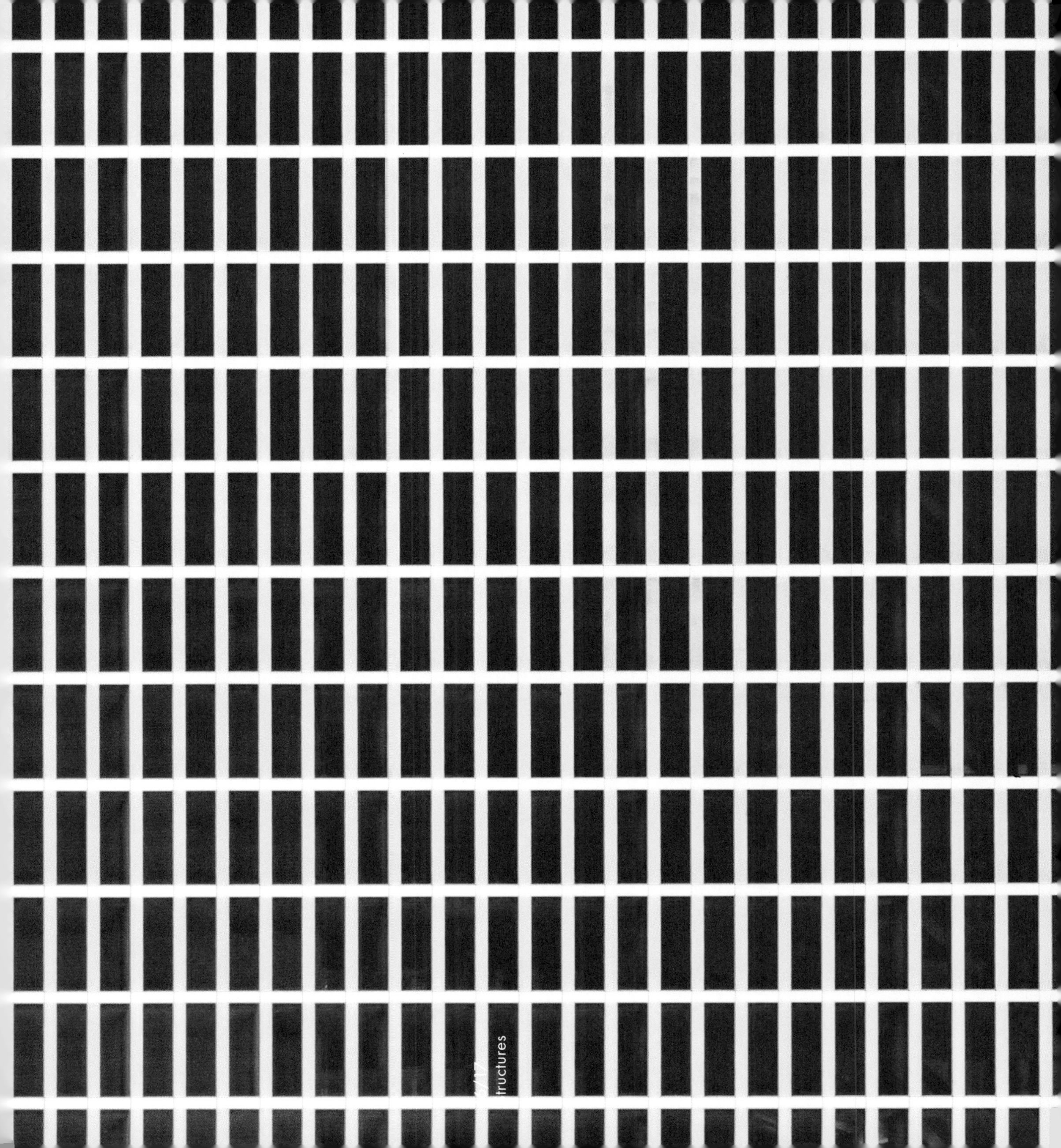
tructures

Four Seasons
Architect: SOM, New York, USA
Planning Time: 2005–2006
Construction time: 2006–2008
Tasks completed: full engineering service except site observation for structural engineering of loadbearing structure and special structures
Client: Four Seasons Hotels and Resorts, Toronto, Canada

The luxurious hotel complex »Four Seasons« in the bay of Bahrain is another impressive verification of the exceptional importance of high-end engineering for the convincing application of demanding designs. The building consists of two towers with a total height of about 290 m, and an adjacent two-story hotel and ballroom complex. The rectangular concrete towers with dimensions of 9 × 36 m form a double core. They are connected with each other by means of the hotel area, as well as a skydeck floating freely at a height of approx. 200 m above the ground. The free span of the skydeck is 22 m. Hotel rooms and luxury apartments are located in the towers. A restaurant and conference rooms are integrated into the skydeck itself.

Auch der luxuriöse Hotelkomplex »Four Seasons« in der Bucht von Bahrain ist ein eindrucksvoller Beleg für die herausragende Bedeutung von High-End-Engineering bei der überzeugenden Umsetzung anspruchsvoller Entwürfe. Das Gebäude besteht aus zwei Türmen mit einer Gesamthöhe von etwa 290 m und einem danebenliegenden zweistöckigen Hotel- und Ballsaalkomplex. Die rechteckigen Betontürme mit Abmessungen von 9 × 36 m bilden einen doppelten Kern. Sie sind mittels des Hotelbereichs sowie des freischwebenden Skydecks in einer Höhe von etwa 200 m über Boden miteinander verbunden. Die freie Spannweite des Skydecks beträgt 22 m. In den Türmen befinden sich Hotelräume und Luxusapartments. Im Skydeck selbst sind ein Restaurant sowie Besprechungsräume untergebracht.

Bangkok Airport
Architect: Murphy/Jahn, Chicago, USA
Planning Time: 1995–2001
Construction time: 2001–2006
Tasks completed: full engineering service except site observation for structural engineering of load-bearing structures, special structures, and facades
Client: New Bangkok International Airport Co. Ltd., Bangkok, Thailand

Airports are a crucial element in modern infrastructure. For millions of people, they provide the first impression of a city or a country. Airports are hubs of international traffic, which handle an impressive number of passengers every day. In order to develop an airport to be an effective, and at the same time splendid, building careful planning is essential right from the outset. An optimal result can be attained only through the close cooperation of all the planners involved. Apart from architectural considerations, in particular questions of the structure and facade planning, as well as energy management, are of decisive importance for the efficiency and attractiveness of an airport.

Over the past 15 years, Werner Sobek has been commissioned with the engineering of airports throughout the world. What all of these projects have in common is their very high design quality, their technical progressiveness and the attention that has been lavished on satisfying local requirements and the interests of the builders alike.

Flughäfen sind ein entscheidendes Element moderner Infrastruktur. Für Millionen von Menschen vermitteln sie den ersten Eindruck einer Stadt oder eines Landes. Flughäfen sind Drehscheiben des internationalen Verkehrs, die Tag für Tag eine beeindruckende Zahl an Passagieren empfangen. Um einen Flughafen zu einem effektiven und zugleich prächtigen Gebäude zu entwickeln, ist von Beginn an eine sorgfältige Planung erforderlich. Nur durch eine enge Zusammenarbeit aller beteiligten Planer ist ein optimales Ergebnis erreichbar. Neben architektonischen Überlegungen sind insbesondere Fragen der Tragwerks- und Fassadenplanung sowie des Energiemanagements von entscheidender Bedeutung für die Effizienz und Attraktivität eines Flughafens.

In den vergangenen 15 Jahren wurde das Büro von Werner Sobek mit dem Engineering von Flughäfen auf der ganzen Welt beauftragt. Allen Projekten gemeinsam ist die sehr hohe Designqualität, ihre technische Fortschrittlichkeit und die Aufmerksamkeit, die dabei den lokalen Bedürfnissen ebenso wie den Interessen der Bauherren gewidmet wurde.

E3
E4
ทางออกขึ้นเครื่อง Gates E1,E1a
ทางออกขึ้นเครื่อง Gates E2,E2a
ทางออกขึ้นเครื่อง Gates E3,5,7,9
ทางออกขึ้นเครื่อง Gates E4,6,8,10

Bangkok Airport

One of the most impressive examples is surely the Suvarnabhumi Airport in Bangkok, up to its completion in the year 2006 the largest construction site in the world. By far the most visible landmark feature of the new airport is a segmental arch roof with dimensions of 582 × 210 m. Resting on just 16 supports, the roof appears to float at a height of 40 m. The individual span widths are 126 m for the main ties and 81 m for the secondary supports. The roof cladding consists of sun-rain protection louvers. Constructed in the same way as aircraft wings, these louvers have a span width of 9 m.

Under the roof is the large, fully-glazed terminal building, with dimensions of 430 × 120 × 40 m. From here, the passengers gain access to the departure gates via a system of flat tubular structures. These concourses have a width of 42 m and an overall length of 3,100 m. They are partially glazed and covered with membranes. For these membrane roofs, a special three-layered, worldwide-patented membrane structure was developed in cooperation with the company Transsolar, the Laboratory for Dynamics and Acoustics and the architects Murphy/Jahn. Despite its delicacy, the membrane has excellent thermal and acoustic qualities, making it perfect for the use at an airport in the tropics.

The enormous spans of the main support girders posed a great challenge in the development of the supporting structure. A minimization of the dead weight and thus maximum structural effectiveness was, of course, required. On the other hand, the supports are highly visible, decisively characterizing the architectural image through their sheer size. Their structural shaping therefore required special attention. Creation-oriented questions, ranging from the welding technology right through to the prefabrication and the lifting of the finished supports onto their pillars (a lifting height of 40 m!) were further parameters influencing the design.

Eines der beeindruckendsten Beispiele ist sicher der Suvarnabhumi Airport in Bangkok, bis zu seiner Fertigstellung im Jahr 2006 die größte Baustelle der Welt. Weithin sichtbares Wahrzeichen des neuen Flughafens ist ein Lamellendach mit Abmessungen von 582 × 210 m. Das Dach ruht auf nur 16 Stützen und scheint dadurch in einer Höhe von 40 m zu schweben. Die einzelnen Spannweiten betragen 126 m für die Hauptbinder und 81 m für die Nebenträger. Die Dachhaut besteht aus Sonnen-Regenschutz-Lamellen, die wie Flugzeugflügel konstruiert sind und 9 m weit spannen.

Bangkok Airport

Unter dem Dach befindet sich das große, vollkommen verglaste Terminalgebäude mit Abmessungen von 430 × 120 × 40 m. Von ihm aus erreicht man die Flugsteige über ein System als »Concourses« bezeichneter flacher Röhren mit einer Breite von jeweils 42 m und einer Gesamtlänge von 3.100 m. Diese Concoursen sind teilweise verglast und zum Teil mit Membranen eingedeckt. Für diese Membran-Dachflächen wurde in Zusammenarbeit mit der Firma Transsolar, dem Laboratorium für Dynamik und Akustik und den Architekten Murphy/Jahn ein spezieller dreilagiger, weltweit patentierter Membranaufbau entwickelt, der die hohen thermischen und akustischen Beanspruchungen abfängt.

Die enormen Spannweiten der Hauptträger stellten bei der Entwicklung der tragenden Konstruktion eine große Herausforderung dar. Natürlich war eine Minimierung des Eigengewichts, also eine maximale tragstrukturelle Effektivität, gefordert. Andererseits sind die Träger weithin sichtbar, sie prägen somit allein schon aufgrund ihrer schieren Größe das architektonische Erscheinungsbild in entscheidender Weise. Ihre Formgebung bedurfte also besonderer Aufmerksamkeit. Herstellungstechnische Fragen von der Schweißtechnik bis hin zur Vorfertigung und dem Hochheben der fertigen Träger auf ihre Stützen – ein Hebevorgang von 40 m Höhe! – waren weitere Parameter, die den Entwurf beeinflussten.

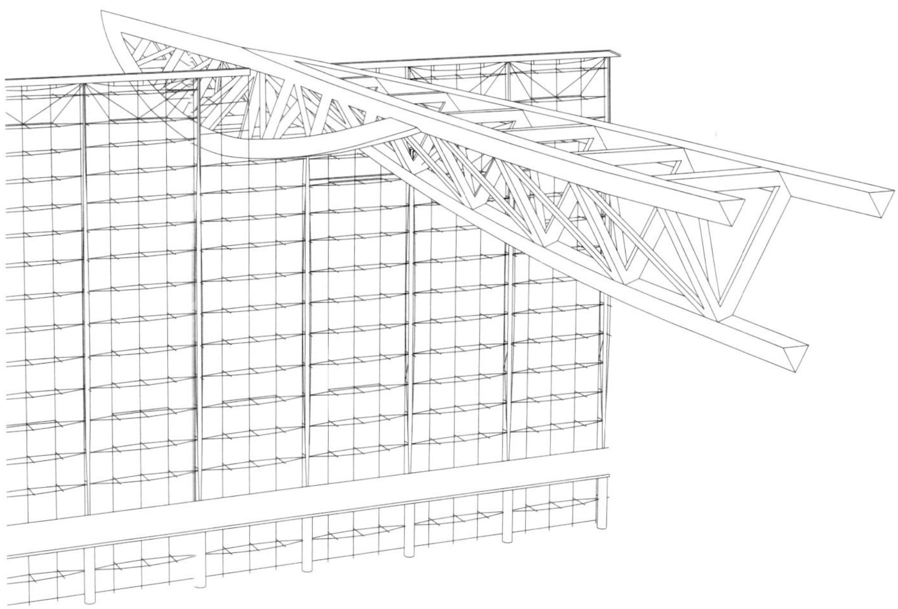

Bangkok Airport

Cologne-Bonn Airport, Terminal 2
Architect: Murphy/Jahn, Chicago, USA
Planning time: 1995 – 1998
Construction time: 1998 – 2000
Tasks completed: full engineering service except site observation for structural engineering of facades and special structures
Client: Flughafen Köln-Bonn GmbH, Germany

The extension of the Cologne-Bonn airport at Terminals 2 and M does not exhibit anything like the dimensional scales of the airport in Bangkok. However, both terminals demonstrate in an impressive manner how close cooperation between architects and engineers can lead to filigree and extremely transparent structures of the highest structural quality. The completely glazed facade of Terminal 2 runs over a length of 800 m around the entire building. The almost dematerialized facade is split into sections, with a length of approx. 40 to 70 m, and a height of between 10 and 20 m. What is special about the facade is its point-anchored insulated glazing that places extremely high demands on the support construction with regard to deformation behavior.

Die Erweiterung des Flughafens Köln-Bonn um die Terminals 2 und M weist bei Weitem nicht die Dimensionen des Flughafens in Bangkok auf. Beide Terminals demonstrieren aber in eindrücklicher Weise, wie eine enge Zusammenarbeit zwischen Architekten und Ingenieuren zu filigranen und äußerst transparenten Strukturen höchster Gestaltungsqualität führen kann. Die vollkommen verglaste Fassade des Terminals 2 verläuft über eine Länge von 800 m um das gesamte Gebäude. Die nahezu entmaterialisierte Fassade ist in Abschnitte mit einer Länge von ca. 40 bis 70 m aufgeteilt und zwischen 10 und 20 m hoch. Ihre Besonderheit besteht in der punktgehaltenen Isolierverglasung, die hinsichtlich des Verformungsverhaltens extrem hohe Anforderungen an die Tragkonstruktion stellt.

D41

Cologne-Bonn Airport, Terminal M
Architect: Murphy/Jahn, Chicago, USA
Planning time: 2003
Construction time: 2003 – 2004
Tasks completed: full engineering service except site observation for structural engineering of load-bearing structure and facades
Client: Flughafen Köln-Bonn GmbH, Germany

The second new departures hall at Cologne-Bonn airport is Terminal M: This consists of filigree arch ties which span the departure level without pillars. The arch ties are supported by means of pendulum supports on a reinforced concrete slab. The horizontal strengthening is guaranteed solely by individual, diagonally-arranged pendulum rods. A further special feature is the extremely transparent facade construction of the departure hall, as well as the passenger and connection bridges.
This transparency is essentially achieved by the slim substructure of the facade structure, as well as by the glass support brackets, which are arranged in the pane joints only.

Die zweite neue Abfertigungshalle am Flughafen Köln-Bonn ist das Terminal M. Dieses besteht aus filigranen Bogenbindern, welche die Abflugebene stützenfrei überspannen und über Pendelstützen auf einem Stahlbetontisch aufgelagert sind. Die horizontale Aussteifung wird nur über einzelne, schräg angeordnete Pendelstäbe gewährleistet. Eine weitere Besonderheit stellt die extrem transparente Fassadenkonstruktion der Abflughalle sowie der Fluggast- und Verbindungsbrücken dar.
Diese Transparenz wird im Wesentlichen durch die schlanke Unterkonstruktion der Fassadenstruktur sowie durch die nur in den Scheibenfugen angeordneten Glashalter erreicht.

Sony Center
Architect: Murphy/Jahn, Chicago, USA
Planning time: 1995–1999
Construction time: 1997–2000
Tasks completed: full engineering service except site observation for structural engineering of special structures and facades
Client: Sony Corporation, Tishman Speyer, Properties Deutschland, Berlin, Germany

The work of Werner Sobek includes numerous office buildings – an especially challenging building type, which combines cost consciousness with demanding architecture and high demands on reliability and user comfort. The Sony Center in Berlin is an impressive example of this combination. Grouped around a central atrium, the building complex is composed of a total of eight separate buildings with different uses (offices and apartments, large movie theaters, a museum of film art etc.). All of the buildings exhibit a large number of special structures, which were a significant prerequisite for the realization of the architectural concept developed by Helmut Jahn. These special structures include the cable facades of up to 20 m in height, spacious conservatories with overhead glazing and large-format glass cubes that encase fragments of the listed pre-war building.

Das Œuvre von Werner Sobek umfasst zahlreiche Bürogebäude – ein besonderer Gebäudetyp, der Kostenbewusstsein mit komplexer Architektur und hohen Ansprüchen an Zuverlässigkeit und Nutzerkomfort kombiniert. Eindrucksvolles Beispiel für diese Kombination ist das Sony Center in Berlin. Es besteht aus einem Gebäudekomplex, der um ein zentrales Atrium gruppiert ist. Der Komplex setzt sich aus insgesamt acht Einzelgebäuden mit unterschiedlichen Nutzungen (Büros und Wohnungen, Großkinos, einem Museum für Filmkunst usw.) zusammen. Alle Gebäude besitzen eine Vielzahl von Sonderkonstruktionen, die wesentliche Voraussetzung für die Realisierung des von Helmut Jahn entwickelten architektonischen Konzeptes waren. Zu diesen »special structures« gehören bis zu 20 m hohe Seilfassaden, weiträumige Wintergärten mit Überkopfverglasung und großformatige Glaskuben, die denkmalgeschützte Fragmente der Vorkriegsbebauung einhüllen.

Bayer Headquarters
Architect: Murphy/Jahn, Chicago, USA
Planning time: 1998–2000
Construction time: 2000–2002
Tasks completed: full engineering service except site observation for structural engineering of load-bearing structure, special structures, and facades
Client: Bayer AG, Leverkusen, Germany

Extremely transparent cable facades are also the distinguishing feature of the new head office of Bayer AG in Leverkusen. An entrance hall surrounded by a cable-stayed glass facade of 15 m in height is attached to the four-story, fully-glazed administration building. The facade is held by vertical steel cables, which are arranged at intervals of approx. 1.50 m. The cables require only relatively low pre-stressing to support the eccentrically connected glass panes of the facade without perceptible deformation. Spring elements at the lower end of the cables ensure that the pre-stressing forces are kept approximately constant, even under varying temperature influences. Under wind load the springs move against a stopper, so that the higher cable stiffness can be fully utilized to limit facade deformation.

Äußerst transparente Seilfassaden prägen auch die neue Hauptverwaltung der Bayer AG in Leverkusen. An das viergeschossige, komplett verglaste Verwaltungsgebäude schließt sich eine Eingangshalle an, die von einer seilstabilisierten Glasfassade von 15 m Höhe umgeben ist. Die Fassade wird durch vertikale Stahlseile gehalten, die in einem Abstand von ca. 1,50 m zueinander angeordnet sind. Die Seile benötigen nur eine relativ geringe Vorspannung, um die exzentrisch angeschlossenen Glasscheiben der Fassade ohne merklich erkennbare Verformungen tragen zu können. Am unteren Ende der Seile befinden sich Federelemente, die dafür Sorge tragen, dass die Vorspannkräfte auch unter Temperatureinflüssen annähernd konstant gehalten werden. Unter Windlast bewegen sich die Federn gegen einen Anschlag, damit die höhere Seilsteifigkeit zur Begrenzung der Fassadenverformungen vollständig genutzt werden kann.

Bayer Headquarters

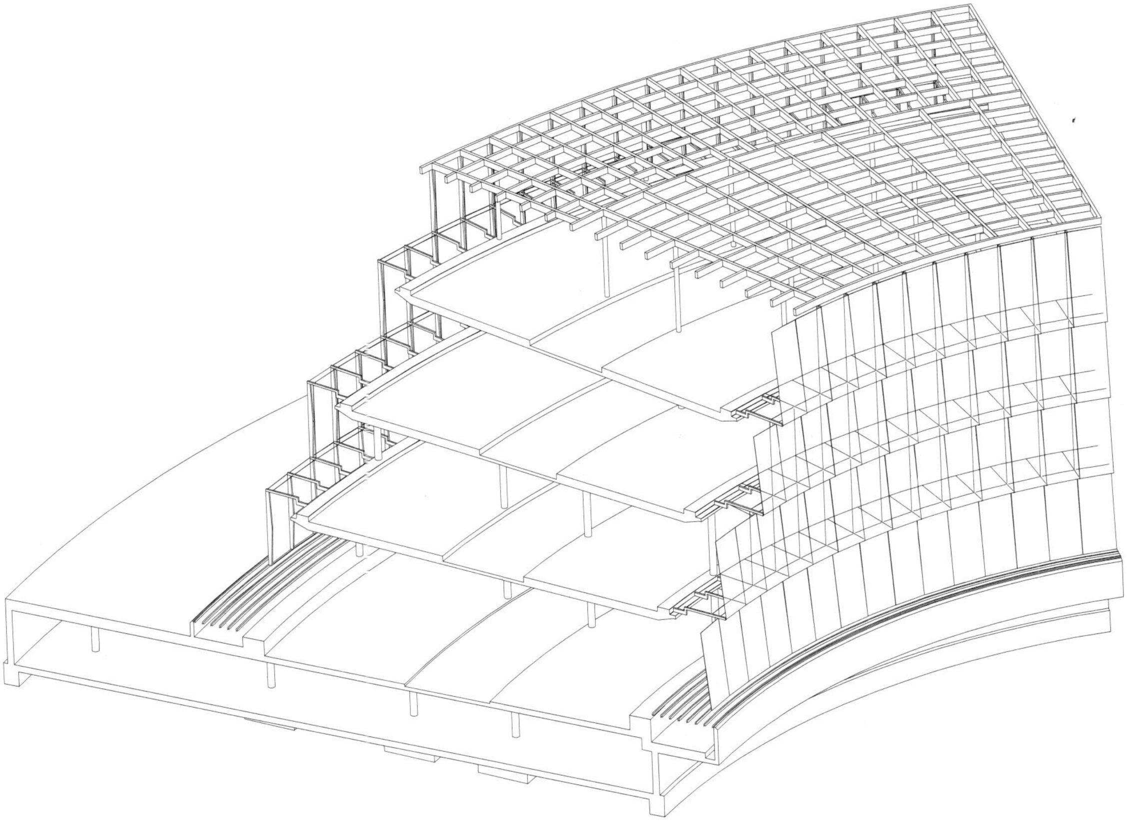

Burda Media Center
Architect: Ingenhoven Architekten,
Düsseldorf, Germany
Planning time: 1997–1999
Construction time: 1999–2000
Tasks completed: full engineering service except
site observation for structural engineering of
loadbearing structure
Client: Burda Immobilien GmbH,
Offenburg, Germany

While the appearance of the Bayer Headquarters is shaped by its facades, the office building of the Hubert Burda Media Holding in Offenburg is characterized by its extraordinary structural form. The transition between wall, floor, ceiling, and facade is fluid and dynamic. This allows a flexible and generous interior design of this bright, cheerful building which does almost entirely without artificial light. The structure is made of reinforced concrete with a curved roof shell. The building itself consists of several »fingers«, between which open spaces with fruit trees are arranged. Every work station offers a view of the meadow landscape. Within the buildings, open stairs generate a room continuum, in which both individual and open-plan offices, group secretarial offices, waiting and discussion areas or archives, can be set up. The room division can be adapted to the changing requirements of editorial work with the aid of system partition walls and flexible furnishings.

Während das Erscheinungsbild der Bayer-Hauptverwaltung insbesondere durch die Fassaden geprägt ist, zeichnet sich das Bürogebäude der Hubert Burda Medien Holding in Offenburg durch sein außergewöhnliches Tragwerk aus. Der Übergang zwischen Wand, Boden, Decke und Fassade ist fließend und dynamisch. Eine flexible und großzügige Innenraumgestaltung ist dadurch ebenso möglich wie der weitgehende Verzicht auf Kunstlicht in diesem hellen, freundlichen Gebäude. Das Tragwerk des Gebäudes besteht aus einer Stahlbetonkonstruktion mit gekrümmter Dachschale. Das Gebäude selbst besteht aus mehreren »Fingern«, zwischen denen sich mit Obstbäumen gestaltete Freiräume befinden. Alle Arbeitsplätze bieten einen Ausblick in die Auenlandschaft. Innerhalb der Gebäude erzeugen offene Treppen ein Raumkontinuum, in dem sowohl Einzel- als auch Kombibüros, Gruppensekretariate, Warte- und Besprechungszonen oder Archive eingerichtet werden können. Die Raumaufteilung kann mithilfe von Systemtrennwänden und einer flexiblen Möblierung den wechselnden Anforderungen redaktioneller Arbeit angepasst werden.

ICE Train Station
Architect: Murphy/Jahn, Chicago, USA
Planning time: 1997 – 2001
Construction time: 2001 – 2002
Tasks completed: full engineering service except site observation for structural engineering of steel structures
Client: Flughafen Köln-Bonn GmbH, Germany

Wide-spanning yet filigree and transparent structures are possible only through the consistent application of the principles of lightweight construction propagated by Werner Sobek. Designing structures which are as light as possible involves testing the limits of what is physically and technically feasible. It is about the aesthetics of the minimum, the search for the unknown and pushing back the limits set by science. Where structures bridge large spans, or achieve great heights, the reduction of dead weight is an economic necessity; frequently it is also a prerequisite for the structure's realizability itself. With rather more mundane constructions involving smaller dimensions, lightweight structures mean savings in the amount of material employed, and in most cases also in the energy employed. It is at this point, at the latest, that the ecological aspect becomes preponderant. Therefore lightweight building is a basic component of the design philosophy of Werner Sobek.

Weitspannende und dennoch filigrane und transparente Tragwerke sind nur durch die konsequente Anwendung der von Werner Sobek dargelegten Prinzipien des Leichtbaus möglich. Das Entwerfen der leichtestmöglichen Konstruktionen ist das Herantasten an das physikalisch und das technisch Machbare. Es hat zu tun mit der Ästhetik des Minimalen, mit der Suche nach dem Unbekannten und mit dem Überschreiten von durch Wissenschaftsdisziplinen gezogenen Grenzen. Bei Konstruktionen, die große Spannweiten überbrücken oder die große Höhen erreichen, ist die Reduktion des Eigengewichtes ökonomischer Zwang, häufig genug auch Voraussetzung für die Realisierbarkeit selbst. Bei den eher alltäglicheren Konstruktionen kleinerer Abmessungen bedeutet Leichtbau eine Ersparnis an eingesetzter Masse, zumeist auch an eingesetzter Energie. Spätestens hier scheint der ökologische Aspekt auf: Leicht zu bauen ist deshalb grundlegender Bestandteil der Entwurfsphilosophie von Werner Sobek.

ICE Train Station

The roof of Bangkok Airport showcased earlier is certainly the most impressive example of large spans. However, numerous other projects, such as the exhibition halls in Hamburg, Munich, or Shanghai, for example, are also characterized by their filigree roof-bearing structure and large spans. If such a structure is combined with glass, a further dimension of lightness results. A vivid example of this is the underground ICE rail station at Cologne-Bonn airport which is over-arched by an almost dematerialized, barrel-shaped roof with dimensions of 156 × 38 m. The roof's supporting structure consists of extremely slim steel arches running in parallel. The span of the arches is 36 m (at an arch height of approx. 3.50 m). These main arches are connected with each other by longitudinal profiles, which are placed at intervals of 1.60 m. Running parallel to the main arches are secondary arches of aluminum. Placed at 2 m intervals, the sole purpose of these secondary arches is to support the glazing.

Eindrucksvollstes Beispiel für große Spannweiten ist sicher das bereits vorgestellte Dach des Flughafens Bangkok. Aber auch zahlreiche andere Projekte wie z.B. die Messehallen in Hamburg, München oder Shanghai zeichnen sich durch ihr filigranes Dachtragwerk und die großen Spannweiten aus. Wird ein solches Tragwerk mit einer Glaseindeckung kombiniert, so ergibt sich eine weitere Dimension der Leichtigkeit. Anschaulich wird dies beim unterirdischen ICE-Bahnhof am Flughafen Köln-Bonn, für den eine nahezu entmaterialisierte tonnenförmige Überdachung von 156 × 38 m entwickelt wurde. Das Tragwerk dieser Überdachung besteht aus parallel verlaufenden, extrem schlanken Stahlbögen. Die Spannweite der Bögen beträgt 36 m (bei einem Bogenstich von ca. 3,50 m). Diese Hauptbögen sind durch Längsprofile, die jeweils im Abstand von 1,60 m stehen, miteinander verbunden. Parallel zu den Hauptbögen verlaufende Nebenbögen aus Aluminium, die jeweils in einem Abstand von 2 m zueinander stehen, dienen nur zur Auflagerung der Verglasung.

Lufthansa Aviation Center
Architect: Ingenhoven Architekten,
Düsseldorf, Germany
Planning time: 1999–2004
Construction time: 2004–2006
Tasks completed: full engineering service except site observation for structural engineering of load-bearing structures, special structures, and facades
Client: MARDU Grundstücks- und Verwaltungsgesellschaft GmbH & Co. KG, Grünwald, Germany

Wide-span, extremely transparent roof structures can also be found in the new head offices of Lufthansa AG at Frankfurt airport. The building has a comb-like structure. There are ten building modules separated by transparent atria. Roofed over with filigree, double-arch steel grating shells, these green garden courtyards serve as acoustic and emission buffers. In order to achieve maximum transparency, the exterior facades of the atria are developed as cable facades: their sole supporting elements are vertically tensioned cables.

Auch in der neuen Hauptverwaltung der Lufthansa AG am Frankfurter Flughafen finden sich solche weit spannenden, äußerst transparenten Dachkonstruktionen. Das Gebäude hat eine kammartige Struktur mit zehn Modulen, zwischen denen sich transparente Atrien befinden. Diese grünen Gartenhöfe, die von filigranen, doppelt gekrümmten Stahlgitterschalen überdacht sind, dienen als Schall- und Immissionspuffer. Um größtmögliche Transparenz zu erreichen, sind die Außenfassaden der Atrien als Seilfassaden ausgebildet, deren einziges Tragelement vertikal gespannte Seile sind.

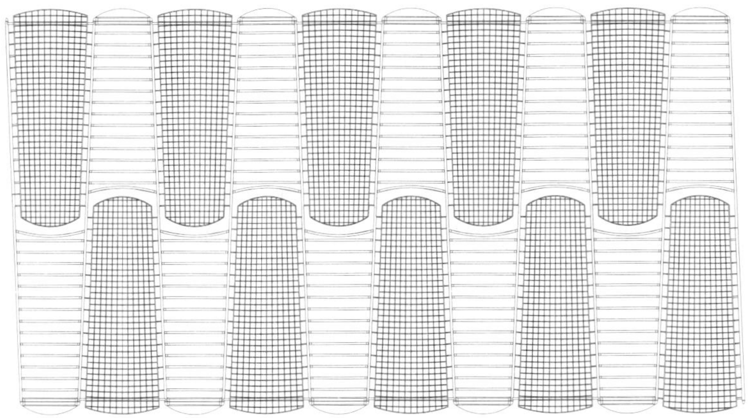

Lufthansa Aviation Center

Charles de Gaulle
Architect: Werner Sobek
Planning time: 2001–2002
Construction time: 2002–2003
Tasks completed: design, object planning and full engineering service
Client: Deutsche Post AG, Bonn, Germany

Bridges are a central component part of our developed environment. Therefore, the objective should never be to resort to a standard solution, but instead to develop a solution appropriate to the basic conditions and to achieve a structure of the highest design quality. This is why in bridge building in particular, the designs of Werner Sobek pay attention not only to an adequate incorporation into the urban setting and to a stringent support structure logic, but also to the development of a uniform structural concept down to the smallest detail.

Brücken sind ein zentraler Bestandteil unserer gebauten Umwelt. Ziel sollte es deshalb nie sein, auf eine Standardlösung zurückzugreifen, sondern eine den Randbedingungen angemessene Lösung zu entwickeln und ein Bauwerk von höchster gestalterischer Qualität zu schaffen. Insbesondere im Brückenbau achten die Entwürfe von Werner Sobek daher nicht nur auf eine angemessene Einbettung in die Stadtlandschaft und eine stringente Tragwerkslogik, sondern auch auf die bis zum kleinsten Detail reichende Entwicklung eines einheitlichen Gestaltungskonzeptes.

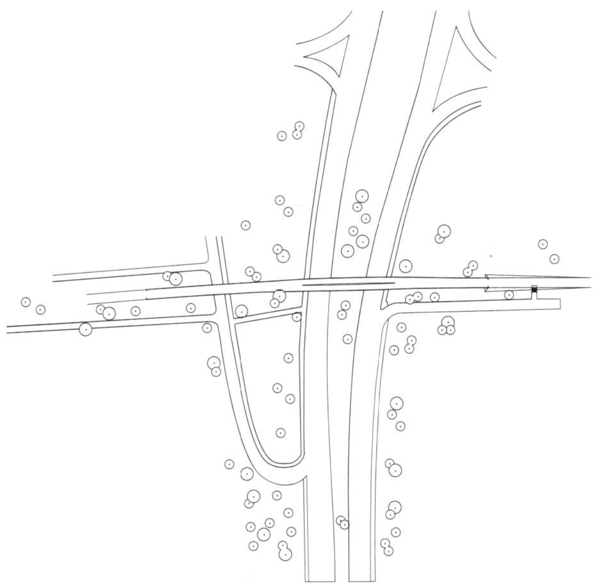

Charles de Gaulle

This design philosophy is very vividly illustrated by the pedestrian bridge over Charles de Gaulle Street in Bonn. The bridge connects the Post Tower with the Rhine meadows. It is thus an important component of the urban integration of the multi-story building and its neighboring buildings. The bridge is 35 m long and 13 m wide. It involves a composite construction with coved steel girders and connecting reinforced concrete plates as a transition to the abutment or to the roofing construction of the entrance gallery. The bridge has a minimum plate thickness of just 120 mm in the field range, giving it its especially slender appearance. Also remarkable is the joint-free bridge construction, which was achieved by a horizontally-elastic mounting of the abutment on piles. The railings are made of curved glass panes.

Sehr anschaulich gemacht wird diese Entwurfsphilosophie durch die Fußgängerbrücke über die Charles-de-Gaulle-Straße in Bonn. Die Brücke verbindet den Post Tower mit den Rheinauen und ist wichtiger Bestandteil der städtebaulichen Einbettung des Hochhauses und seiner Nebenbauten. Sie ist 35 m lang und 13 m breit. Es handelt sich um eine Verbundkonstruktion mit gevouteten Stahlträgern mit anschließenden Betonvollplatten als Übergang zum Widerlager bzw. zur Überdachungskonstruktion der Zufahrtsgalerie. Die Brücke verfügt über eine minimale Plattenstärke von nur 120 mm im Feldbereich und ist dadurch in der Ansicht besonders schlank. Bemerkenswert ist auch die fugenlose Brückenkonstruktion, die durch eine horizontal elastische Lagerung des Widerlagers auf Pfählen erreicht werden konnte. Das Geländer besteht aus gekrümmten Glasscheiben.

The pedestrian bridges in Dortmund are a further example of elegant and clear traffic construction works. The bridges cross two multi-lane traffic routes (in a prominent urban position) and are 432 m and 296 m long respectively. They are elegantly curved and branch out at one point. The elegant forms, in combination with unusual support structure concepts, create the distinctive appearance that the builder wished to achieve. The main parts of both bridges are wrought in steel. The subordinate bays, including the slim piers which are placed close together to support the bridge construction, are made of concrete. So as not to interrupt the harmonious appearance of the bridges with lamp posts, a lighting concept was developed involving the integration of LEDs into the hand rail.

Weitere Beispiele für elegante und klare Verkehrsbauwerke sind die Fußgängerbrücken in Dortmund, die – an städtebaulich herausragender Position – zwei mehrspurige Verkehrswege überqueren. Die Bauwerke sind 432 m beziehungsweise 296 m lang. Die Brücken sind im Grundriss geschwungen und verzweigen sich an einer Stelle. Die harmonischen Formen in Verbindung mit nicht alltäglichen Tragwerkskonzepten bewirken die vom Bauherrn gewünschte Prägnanz des Erscheinungsbildes. Die Hauptfelder sind bei beiden Brücken in Stahlbauweise gefertigt. Die nachgeordneten Brückenfelder einschließlich der schlanken Pfeiler, die die Brückenbauwerke in engen Abständen stützen, wurden in Betonbauweise ausgeführt. Um das Erscheinungsbild der Brücken nicht durch Pfeiler zu beeinträchtigen, wurde ein Beleuchtungskonzept entwickelt, das auf in den Handlauf integrierten LEDs beruht.

Victor Toyka
Architect: Werner Sobek
Planning time: 2000–2004
Construction time: 2004–2006
Tasks completed: design, object planning and full engineering service
Client: City of Dortmund, Germany

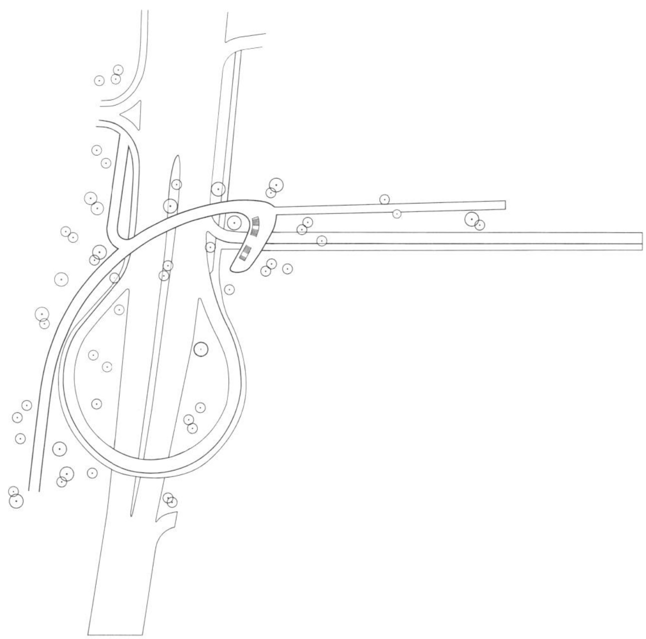

Waldschlösschenbrücke
Architect: Werner Sobek
Planning time: 2007 (competition, 1st prize)
Tasks completed: overall design
Client: City of Dresden, Germany

An even more extensive dematerialization and discreet integration into the urban environment can be found in the prize-winning design for the Waldschlösschenbrücke in Dresden. Here, in particular, it was a matter of renouncing importance and abandoning anything that was technically imposing along with the quest for record span widths. By concentrating on the minimum and on inconspicuousness, a design of the highest aesthetic quality was produced. The total width of the narrow and permeable bridge body is a mere 19 m. The installation height can be limited to a maximum of 1.35 m through a continuous, narrow support. The pillars themselves are only lightly loaded and thus can also be designed to be very slim. The very light bridge body has a longitudinal gloss line, which reduces its visually perceived size even further. As a result of this, the bridge seems to float, an impression which conveys a feeling of lightness to pedestrians, to playing children or to hikers who experience the bridge from below and in direct proximity.

Eine noch weiter reichende Entmaterialisierung und diskrete Einfügung in das städtebauliche Umfeld findet sich beim preisgekrönten Entwurf für die Waldschlösschenbrücke in Dresden. Hier ging es insbesondere um das Zurücknehmen von Wichtigkeit, den Verzicht auf das technisch Imposante und auf die Jagd nach Rekordspannweiten. Durch Arbeit am Minimalen, am Unscheinbaren entstand ein Entwurf von höchster ästhetischer Qualität. Die Gesamtbreite des schmal und durchlässig angelegten Brückenkörpers beträgt lediglich 19 m. Die Bauhöhe kann durch eine kontinuierliche, enge Stützung auf maximal 1,35 m beschränkt werden. Die Stützen selbst sind jeweils nur geringfügig belastet und können somit ebenfalls sehr schlank ausgeführt werden. Der sehr hell gehaltene Brückenkörper besitzt eine längslaufende Glanzlinie, die seine optisch wahrgenommene Größe weiter reduziert. Die Brücke beginnt hierdurch förmlich zu schweben, ein Eindruck, der auch dem Fußgänger, den spielenden Kindern oder den Wanderern, welche die Brücke von unten und aus unmittelbarer Nähe erleben, ein Gefühl von Leichtigkeit vermittelt.

Audi Motorshow
Architect: Ingenhoven Architekten, Düsseldorf, Germany
Planning time: 1998–1999
Construction time: 1999–2000
Tasks completed: full engineering service except site observation for structural engineering of loadbearing structure and special structures
Client: Audi AG, Ingolstadt, Germany

Besides classical constructions, such as airports, office buildings, and bridges, Werner Sobek also bears responsibility for the engineering of numerous other structures. An example of this is the trade fair stand of Audi AG, which was planned jointly with the Dusseldorf architect Christoph Ingenhoven. The structure of these trade fair stands, which were used at all large-scale automobile trade fairs, consisted of a glazed, repeatedly-curving wave. The wave, which was used as a screen for the projection of colored light and changing film sequences, was 6 m high and, depending on stand size, between 100 and 300 m long. The 4 mm thick glass panes were held by a fine-meshed cable network, which was tightened into a supporting rod structure composed of arched, crossing stainless-steel pipes. Apart from the extremely complex geometry, the engineers also had to consider that the stand had to be assembled and dismantled in a very short space of time; moreover, it had to be reusable several times – a requirement which made this project one of the most difficult challenges facing the office, however, one which also resulted in a wonderfully light, perfectly formed structure.

Neben klassischen Bauten wie Flughäfen, Bürogebäuden und Brücken zeichnet Werner Sobek auch für das Engineering zahlreicher anderer Strukturen verantwortlich. Ein Beispiel hierfür ist der Messestand der Audi AG, der gemeinsam mit dem Architekten Christoph Ingenhoven aus Düsseldorf geplant wurde. Die Struktur dieser auf allen großen Automobilmessen eingesetzten Messestände basierte auf einer gläsernen, mehrfach gekrümmten Welle, die durch die Projektion von farbigem Licht und wechselnden Filmsequenzen atmosphärisch in Szene gesetzt wurde. Die Wand war 6 m hoch und je nach Standgröße 100 bis 300 m lang. Die 4 mm starken Scheiben aus satiniertem Glas wurden von einem engmaschigen Seilnetz gehalten, das in eine tragende Stabstruktur aus gebogenen, sich kreuzenden Edelstahlrohren gespannt wurde. Neben der äußerst komplexen Geometrie mussten die Ingenieure auch berücksichtigen, dass der Stand in sehr kurzer Zeit auf- und abbaubar und mehrmals wieder verwendbar sein musste – eine Anforderung, die dieses Projekt zu einer der schwierigsten Herausforderungen für das Büro machte, aber auch zu einer wunderbar leichten, im Detail ebenso wie im Ganzen formvollendeten Struktur führte.

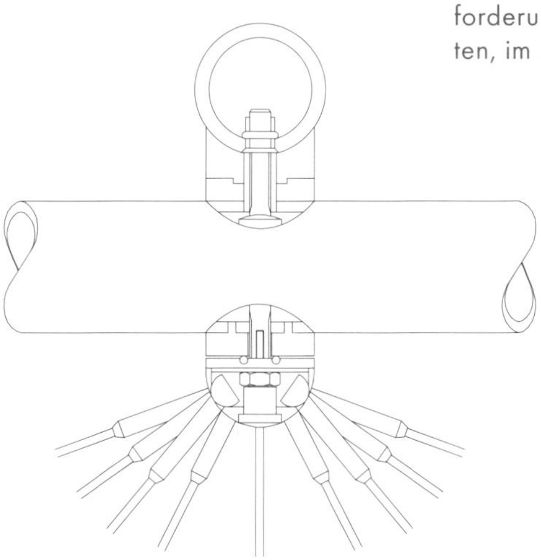

Audi Motorshow

Burda Car Park
Architect: Ingenhoven Architekten, Düsseldorf, Germany
Planning time: 1999–2001
Construction time: 2001–2002
Tasks completed: full engineering service except site observation for structural engineering of loadbearing structure and special structures
Client: Burda Immobilien GmbH, Offenburg, Germany

The circular-shaped multi-story car park of the Burda Media Park in Offenburg was also the product of collaboration with Christoph Ingenhoven. A semi-transparent cable facade lies over the building like a veil. The facade consists of pre-assembled cable elements onto which prepared round wooden elements were attached using system brackets. Two steel rings are used for weight, and at the same time incorporate the lighting. The result of this simple and low-cost yet high-quality implementation, are visual references which continuously alternate between inside and outside, as well as a very harmonious overall appearance. The compact construction method of the naturally-ventilated multi-story car park also requires a special mention. Via the interior ascent ramps and the stairwell, all vertical openings are integrated into the overall volume.

Auch das kreisrunde Parkhaus des Burda Medienparks in Offenburg entstand aus der Kooperation mit Christoph Ingenhoven. Über dem Gebäude liegt wie ein Schleier eine semitransparente Seilfassade. Diese Fassade besteht aus vorgefertigten Seilelementen, an die über Systemhalter Rundhölzer gehängt wurden. Zwei Stahlringe dienen der Ballastierung und nehmen zugleich die Beleuchtung auf. Das Ergebnis dieser einfachen und kostengünstigen, aber dennoch hochwertigen Ausführung sind immer wieder wechselnde Blickbezüge zwischen innen und außen sowie ein sehr harmonisches Erscheinungsbild. Neben der außergewöhnlichen Fassadenkonstruktion ist auch die kompakte Bauweise des natürlich belüfteten Parkhauses hervorzuheben. Durch die innenliegenden Auffahrtrampen und das Treppenhaus sind alle vertikalen Erschließungen in das Gesamtvolumen integriert.

Merck Serono
Architect: Murphy/Jahn, Chicago, USA
Planning time: 2002–2004
Construction time: 2004–2006
Tasks completed: full engineering service except site observation for structural engineering of load-bearing structure, special structures, and facades
Client: Merck Serono International S.A., Geneva, Switzerland

The new head office of the company Merck Serono in Geneva is another example of extraordinary support structures. Besides numerous special structures, such as a horizontally pre-stressed cable facade, a scaled single facade or wide-spanning bridge constructions on the inside, the main feature is the 1,000 m^2 roof over the forum. The delicate, extremely transparent steel-glass construction can be opened with the aid of several hydraulic cylinders, thus allowing the simple and rapid natural ventilation of the entire interior space. It is the largest openable glass construction in the world.

Ein weiteres Beispiel für außergewöhnliche Tragwerke ist die neue Hauptverwaltung der Firma Merck Serono in Genf. Neben zahlreichen Sonderkonstruktionen wie einer horizontal vorgespannten Seilfassade, einer geschuppten Einfachfassade oder weit spannenden Brückenkonstruktionen im Inneren ist v.a. das 1.000 m^2 große Dach über dem so genannten Forum bemerkenswert: Die grazile, äußerst transparente Stahl-Glas-Konstruktion kann mithilfe mehrerer Hydraulikzylinder geöffnet werden und erlaubt so eine einfache und rasche natürliche Belüftung des gesamten Innenraums. Dieses Dach ist die größte öffenbare Glaskonstruktion der Welt.

Merck-Serono

Skins

The primary objective of an envelope is protection. The envelope protects an object against external influences, such as heat, cold, wind, moisture or light, for example. Clothing is the second skin of a person – building shells are literally its third skin. That is why this chapter is entitled »Skins« – because building envelopes are more than just a space closure. The facades presented here are highly complex, multi-functional structures and represent both the face and clothing of a building in one. Visual, thermal, and acoustic qualities are accorded equal consideration in their planning. Besides steel and glass, numerous other materials, such as titanium, textiles, sandwich elements or wood are used. The following pages can give only a very small insight into the variety of these building envelopes.

Der primäre Zweck einer Hülle ist Schutz: Die Hülle schützt ein Objekt vor äußeren Einflüssen wie z.B. Hitze, Kälte, Wind, Feuchtigkeit oder Licht. Bekleidung ist die zweite Haut des Menschen – Gebäudehüllen sind im wahrsten Sinne des Wortes seine dritte Haut. Aus diesem Grund lautet die Überschrift dieses Kapitels auch »Skins« – denn Gebäudehüllen sind mehr als nur ein Raumabschluss. Die hier vorgestellten Fassaden sind hochkomplexe, multifunktionale Strukturen, Gesicht und Kleid eines Gebäudes in einem. Optische, thermische und akustische Qualitäten werden bei ihrer Planung gleichermaßen berücksichtigt. Neben Stahl und Glas kommen zahlreiche andere Materialien wie Titan, Textilien, Sandwichelemente oder Holz zum Einsatz. Die folgenden Seiten können dabei nur einen kleinen Einblick in die Vielfalt dieser Gebäudehüllen geben.

A transparent envelope that protects us against the elements, while at the same time allowing direct visual contact with the surrounding nature – the gap on this ideal is steadily being closed in the work presented here through a constantly advancing process of dematerialization. The view outside broadens horizons, not only visually, but also intellectually and psychologically. A supply of natural light is one of the basic needs of the human body. Transparency is therefore a deciding factor in the work presented here. The most important material in achieving this objective is glass – a fascinating building material, whose areas of application have been extended continuously in recent years, not least by the research work of Werner Sobek.

Eine transparente Hülle, die den Menschen vor den Unbilden des Wetters wie Regen, Wind oder Kälte schützt, ihm gleichzeitig aber einen direkten visuellen Kontakt mit der umgebenden Natur erlaubt – dieser Idealvorstellung kommen die hier gezeigten Arbeiten durch eine stetig voranschreitende Entmaterialisierung immer näher. Der Blick nach außen weitet die Wahrnehmung nicht nur in optischer, sondern auch in geistiger und in seelischer Hinsicht. Die Versorgung mit natürlichem Licht ist eines der Grundbedürfnisse des menschlichen Körpers. Deshalb ist Transparenz ein entscheidender Faktor der vorgestellten Arbeiten. Wichtigster Werkstoff für das Erreichen dieses Ziels ist Glas – ein faszinierendes Material, dessen Einsatzgebiete nicht zuletzt durch die Forschungen von Werner Sobek in den letzten Jahren ständig erweitert wurden.

Rhön Clinic
Architect: Lamm, Weber, Donath & Partner, Stuttgart, Germany
Planning time: 1996
Construction time: 1996–1997
Tasks completed: full engineering service except site observation for structural engineering of total structure
Client: Rhön-Klinikum AG, Bad Neustadt, Germany

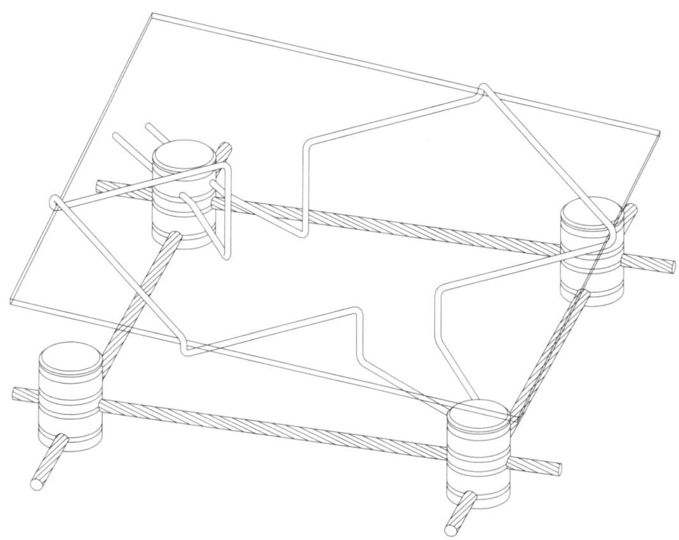

Rhön Clinic

The overall aim of Werner Sobek is to create a transparent envelope that protects the user without isolating him from his environment. There are various possible solutions to this problem, all very different from traditional support structure solutions. This is illustrated especially vividly by the coverage of the outdoor facilities of the Rhön clinic in Bad Neustadt. In order to enable the patients of this specialist heart clinic to access the external area without being exposed to the elements, large areas of the park have been covered over with a glazed cable network. The cable network is raised through timber pillars and tensioned with cables. The covering of the network is implemented through a specially developed glass-shingle system, which combines the advantages of a cable network construction with those of real glazed roofing. Standardized glass shingles can thus be attached within a short space of time to form a network, whose grid meshes have very different geometry. The result is a filigree envelope of high aesthetic quality that sways gently in the breeze.

Die möglichen Lösungen für eine transparente Hülle, die den Menschen schützt, ihn aber nicht von seiner Umgebung abtrennt, sind sehr verschiedener Natur und unterscheiden sich deutlich von traditionellen Tragwerksvorstellungen. Besonders anschaulich wird dies bei der Überdeckung der Außenanlagen der Rhönklinik in Bad Neustadt. Um den Patienten dieser auf komplexe Herzoperationen spezialisierten Klinik einen vor Wind und Wetter geschützten Zugang zum Außenbereich zu ermöglichen, wurden große Teile der Grünanlagen mit einem verglasten Seilnetz überdeckt. Dieses Seilnetz ist durch Holzstützen aufgeständert und mit Seilen abgespannt. Die Eindeckung des Netzes erfolgt durch ein speziell entwickeltes Glasschindelsystem, das die Vorteile einer Seilnetzkonstruktion mit denen einer Echtglasüberdachung verbindet. Standardisierte Glasschindeln können so innerhalb kurzer Zeit auf ein Netz aufgebracht werden, dessen Maschen sehr unterschiedliche Geometrien haben. Das Ergebnis ist eine filigrane, sich leicht im Wind wiegende Hülle von hoher ästhetischer Qualität.

Bremen University
Architect: Jan Störmer, Hamburg, Germany
Planning time: 1997–2000
Construction time: 1998–2000
Tasks completed: full engineering service except site observation for structural engineering of load-bearing structure, special structures, and facades
Client: Senator für Bildung und Wissenschaft, Kunst und Sport der Hansestadt Bremen, Germany

The new entrance area of Bremen University is also an almost entirely dematerialized, highly transparent building envelope. The support construction for the facades consists solely of vertically tightened cables held by springs. The glass hall has a floor area of 22 × 43.50 m and is 15 m high. The primary structure of the hall is a steel grid, which is supported at its front longitudinal side by six V-stays, at the rear longitudinal side by six pendulum supports attached to the existing buildings. Reinforcement is provided by a cable support of the external V-stays and through wind braces between the pendulum supports. The cable facade can bend with the wind by up to 350 mm, both towards the inside and the outside. This construction was the first facade in the world to be held exclusively by vertically pre-stressed cables, and marked an important step towards maximum transparency in a building shell.

Auch der neue Eingangsbereich der Universität Bremen ist eine nahezu entmaterialisierte Gebäudehülle von höchster Transparenz. Als Tragkonstruktion für die Fassaden dienen einzig vertikal gespannte Seile, die in Federn gelagert sind. Die Glashalle hat eine Grundfläche von 22 × 43,50 m und ist 15 m hoch. Das Primärtragwerk der Halle besteht aus einem Trägerrost, der an seiner vorderen Längsseite von sechs V-Stützen, an der hinteren Längsseite von sechs auf den vorhandenen Gebäuden aufstehenden Pendelstützen getragen wird. Die Aussteifung erfolgt durch eine Seilabspannung der jeweils äußeren V-Stützen und durch Windverbände zwischen den Pendelstützen. Die Seilfassade kann sich unter Windeinwirkung um bis zu 350 mm – nach innen wie nach außen – durchbiegen. Diese Konstruktion war die erste ausschließlich durch vertikal vorgespannte Seile gehaltene Fassade der Welt und markierte einen wichtigen Schritt hin zur größtmöglichen Transparenz in einer Gebäudehülle.

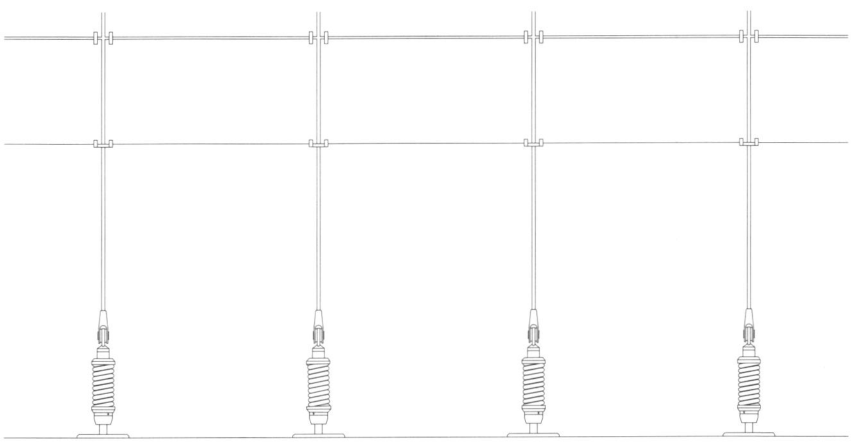

SWM Reutlingen
Architect: Allmann, Sattler, Wappner,
Munich, Germany
Planning time: 1998–2000
Construction time: 2000–2001
Tasks completed: full engineering service except site observation for structural engineering of loadbearing structure
Client: Verband der Metall- und Elektroindustrie Baden-Württemberg, Germany

Revolutionary new facade systems allow an unprecedented degree of transparency in modern architecture. Depending on user requirements and the architectural design, however, facades and building shells can also serve to deliberately block off views from the outside or to achieve a precise type of visual relationship. Metal can play an important role here – whether as a surface element, a perforated plate, a woven fabric or even as a tubular construction. An especially vivid example of this interplay with light and materials is the SWM administration building in Reutlingen. The complex was designed for Südwestmetall, an association of entrepreneurs working in the metal sector. It consists of three buildings almost completely covered with an envelope of stainless steel. On the ground floor, embossed panels are a source of ornamentation for the facade and of natural lighting for the interior – without really allowing a full view of the inside. The facade thus allows those inside to look out, but prevents those on the outside from looking in. On the upper floors, a fine perforation of the plates allows light to enter in the window area, without breaking up the uniform appearance of the secondary steel facade.

Neue Fassadensysteme erlauben einen bisher nicht gekannten Grad an Transparenz. Abhängig von Nutzeranforderung und architektonischem Entwurf können Fassaden und Gebäudehüllen aber auch dazu dienen, Einblick zu verwehren oder eine präzise Art von Blickbeziehung zu schaffen. Ein wichtiges Instrument hierfür ist Metall – sei es als flächiges Element, Lochblech, Gewebe oder auch als Röhrenkonstruktion. Besonders anschaulich zeigt sich dieses Spiel mit Licht und Materialien am Verwaltungsgebäude des Arbeitgeberverbandes Südwestmetall in Reutlingen. Der aus drei Gebäuden bestehende Komplex ist nahezu komplett mit einer Haut aus Edelstahl verkleidet. Im Bereich des Erdgeschosses sorgen gestanzte Bleche für eine Ornamentierung der Fassade und eine natürliche Beleuchtung der Innenräume – ohne jedoch den Blick nach innen wirklich zu öffnen. Die Fassade gewährt so zwar einen Ausblick, aber keinen Einblick. In den oberen Geschossen sorgt eine feine Perforierung der Bleche im Fensterbereich für Lichteinfall, ohne das homogene Erscheinungsbild der metallischen Sekundärfassade zu durchbrechen.

Trumpf Entrance
Architect: Barkow Leibinger, Berlin, Germany
Planning time: 2005–2006
Construction time: 2007
Tasks completed: full engineering service except site observation for structural engineering of load-bearing structure, special structures, and facades
Client: Trumpf GmbH & Co. KG, Ditzingen, Germany

The entrance pavilion at the Trumpf factory in Ditzingen represents an alternative form of extreme dematerialization. Central element of this highly transparent facade is an acrylic fin. The fin is linked to the glass panes of the facade and serves as a bracing post. Thanks to the use of acrylic, the posts attain a slenderness that far exceeds the technical limits imposed on conventional glass fins. The transparency and reduction of the structure is further underlined by the extremely slender canopy cantilevering nearly 16 m.

Der Eingangspavillon der Firma Trumpf in Ditzingen zeigt eine alternative Form der extremen Entmaterialisierung. Zentrales Element dieser hochtransparenten Fassade ist ein Acrylglas-Schwert, das als aussteifender Pfosten mit den Fassadenscheiben verbunden ist. Durch den Einsatz von Acrylglas konnten die Pfosten in einer Schlankheit gefertigt werden, die weit jenseits der produktionstechnischen Beschränkungen für herkömmliche Glasschwerter liegt. Transparenz und Reduzierung dieser Konstruktion werden zusätzlich durch das extrem schlanke Vordach unterstrichen, das fast 16 m frei auskragt.

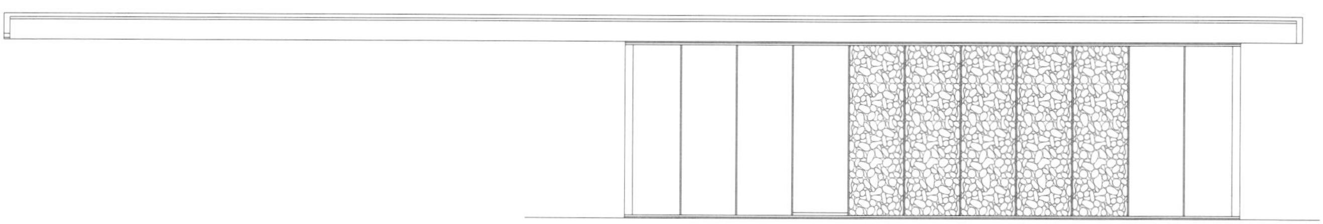

Alongside metal and glass, textiles are among the most important materials for the design of innovative building envelopes. Their spatial flexibility is, beside their low weight and their translucence, the greatest advantage of membranes. An exceptional example, which ideally illustrates all three of these properties of textiles, is the roofing of the center court at Rothenbaum Stadium in Hamburg.

Neben Metall und Glas sind Textilien sicher die wichtigsten Materialien für innovative Gebäudehüllen. Größter Vorteil von Membranen ist neben ihrem geringen Gewicht und ihrer Transluzenz die räumliche Flexibilität. Ein herausragendes Beispiel, das alle drei dieser Eigenschaften von Textilien in idealer Weise nutzt, ist die Überdachung des Center Court am Rothenbaum in Hamburg.

Rothenbaum
Architect: ASP Schweger, Hamburg, Germany
Planning time: 1993–1997
Construction time: 1997–1999
Tasks completed: conceptual design and full engineering service for structural engineering of the roof structure
Client: Deutscher Tennisbund Rothenbaum Turnier GmbH, Hamburg, Germany

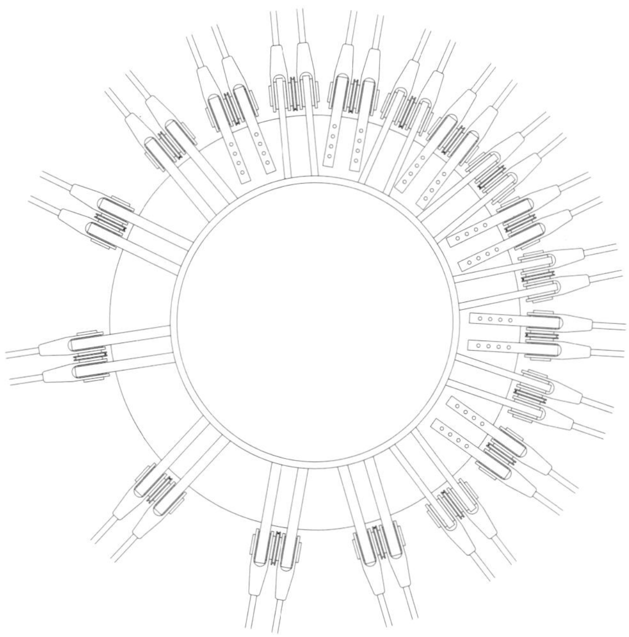

Rothenbaum

The roof spans the stands and the courts with a free span of approx. 102 m. The membrane structure, which was derived from the design principle of the spoked wheel, has an external, permanently-roofed part and a movable inside roof with a diameter of approx. 63 m. With an openable surface area of 3,000 m², Rothenbaum thus boasts one of the largest folding roofs in the world which is usable all year round. The movement of the roof is sensor-controlled and is implemented in a fully automated manner at the push of a button. An integrated meteorological station extrapolates the weather data and analyzes the best time for opening and closing. The spectators can remain seated during the opening and closing procedure – which takes only a few minutes.
Das Dach überspannt Tribünen und Spielfläche mit einer freien Spannweite von ca. 102 m. Die Membranstruktur, die aus dem Konstruktionsprinzip des Speichenrades abgeleitet wurde, besitzt einen äußeren, permanent bedachten Teil und ein bewegliches Innendach mit ca. 63 m Durchmesser. Dies entspricht einer zu öffnenden Fläche von 3.000 m² und macht Rothenbaum somit zum größten ganzjährig nutzbaren Faltdach der Welt. Die Bewegung des Daches ist sensorgesteuert und geschieht vollautomatisiert auf Knopfdruck; eine integrierte meteorologische Station extrapoliert die Wetterdaten und analysiert den besten Zeitpunkt für das Öffnen und Schließen des Daches. Die Zuschauer können während des Auf- und Zufahrens – das jeweils nur wenige Minuten dauert – sitzen bleiben.

Rothenbaum

Station Z
Architect: hg Merz, Berlin, Germany
Planning time: 1999–2003
Construction time: 2003–2005
Tasks completed: full engineering service except site observation for membrane structure
Client: Liegenschafts- und Bauamt Bernau, Germany

For Rothenbaum, it was the mobility and lightness of membranes that were of primary importance. For Station Z, on the other hand, the visual properties of textiles played a determining role. Ground relics are all that survive of the crematorium of the Sachsenhausen concentration camp. In order to protect these traces of the past, an unobtrusive but efficient protective structure had to be developed. The result is a column-free steel structure covering about 2,000 square meters whose uniform surface shimmers translucently. To achieve this effect, the roofing was developed and implemented as a low-air-pressure stabilized membrane construction – a concept that was used for the first time worldwide and is as such a milestone in textile architecture. By using an extremely translucent, PTFE-coated fiberglass membrane and specifically adapted steel grating, a uniform luminance was achieved leaving the visitor to the memorial guessing about the interior support structure. An important factor for the appearance, besides the detailing and the finish, is the apparent levitation of the entire construction above the ground. In order to interfere as little as possible with the substance of the memorial, the protecting envelope rests on just eight small foundation bases which also house all installations and the air suction units.

Während für Rothenbaum v.a. die Beweglichkeit und Leichtigkeit von Membranen ausschlaggebend war, spielten beim Entwurf für Station Z die optischen Eigenschaften von Textilien eine entscheidende Rolle. Von dem Krematorium des Konzentrationslagers Sachsenhausen existieren nur noch Bodenrelikte. Um diese Spuren der Vergangenheit zu erhalten, wurde ein Schutzbau errichtet, der als eine transluzent durchscheinende Hüllkonstruktion mit einer homogenen Oberfläche in Erscheinung tritt. Diese spezielle Überdachung wurde als unterdruckstabilisierte Membrankonstruktion entwickelt und ausgeführt – ein Konzept, das hier weltweit erstmalig zum Einsatz kam und damit einen Meilenstein in der textilen Architektur darstellt. Durch die Verwendung einer äußerst transluzenten, PTFE-beschichteten Glasfasermembrane und spezifisch angepasster Stahlgitterroste wurde eine homogene Leuchtdichte erreicht, welche die innenliegende Tragstruktur für den Besucher der Gedenkstätte nur noch erahnen lässt. Ein wichtiger Faktor für das Erscheinungsbild ist neben der Detaillierung und der Konfektion auch das scheinbare Schweben der gesamten Konstruktion über dem Boden. Um möglichst wenig in die Substanz des Mahnmals einzugreifen, lagert die schützende Hülle lediglich auf acht kleinen Fundamentsockeln. In diesen Sockeln sind die Installationen wie auch die Luftabsaugung untergebracht.

syn chron
Artist: Carsten Nicolai, Berlin, Germany
Architect: LIN Finn Geipel Giulia Andi, Berlin, Germany
Planning time: 2004–2005
Construction time: 2005
Tasks completed: full engineering service except site observation for structural engineering of total structure
Client: Freunde guter Musik Berlin e.V., Germany

Many »skins« cannot be clearly categorized, crossing the boundaries between different crafts, materials, and functions. An example of this is the walk-in sculpture »syn chron«, designed by Berlin artist Carsten Nicolai for the New National Gallery in Berlin. The envelope of this sculpture is used not only to confine a space, but also as resonance body and projection surface. Electronic music, specially composed by Carsten Nicolai, generates laser light modulations on the translucent skin. The architectural body thus becomes a sound and light space at the same time, which the visitor can enter or observe from the outside. The crystal-shaped construction consists of a spatial, irregular, polygonal hollow profile frame without right angles. To allow the sculpture to be assembled repeatedly at different exhibition locations, the structure had to be designed in such a way that it can be completely disassembled. To this end, a special construction was developed of epoxy sandwich panels with aluminum honeycomb core, which were fixed by means of a binder zip – materials and methods never previously employed in construction in this form.

Viele »Skins« lassen sich nicht eindeutig einer bestimmten Kategorie zuordnen, sondern stehen im Grenzbereich verschiedener Gewerke, Materialien und Funktionen. Ein Beispiel hierfür ist die begehbare Skulptur »syn chron«, die der Berliner Künstler Carsten Nicolai für die Neue Nationalgalerie in Berlin konzipiert hat. Die Hülle dieser Skulptur dient nicht nur dem Raumabschluss, sondern auch als Resonanzraum und Projektionsoberfläche. Eine eigens von Carsten Nicolai komponierte elektronische Musik generiert auf der transluzenten Haut Laserlichtmodulationen. Der architektonische Korpus wird somit zugleich Klang- und Lichtraum, in den die Besucher eintreten, den sie aber auch von außen wahrnehmen können. Die kristallförmige Konstruktion besteht aus einem räumlichen, unregelmäßig polygonalen Hohlprofilrahmen ohne rechte Winkel. Um einen mehrfachen Aufbau an verschiedenen Ausstellungsorten zu ermöglichen, musste die Struktur so konstruiert werden, dass sie komplett demontierbar ist. Hierfür wurde eine spezielle Konstruktion aus Epoxysandwichplatten mit Aluminiumwabenkern entwickelt, die durch Klettverschlüsse fixiert wurden – Materialien und Konstruktionsmethoden, die in dieser Form bisher noch nie im Bauwesen angewandt worden waren.

syn chron

Polylit
Artist: Carsten Nicolai, Berlin, Germany
Planning time: 2006
Construction time: 2006
Tasks completed: full engineering service except site observation for structural engineering of load-bearing structure and facades
Client: Kunstmuseum Stuttgart, Germany

The sculpture »polylit« at the Kleiner Schlossplatz in Stuttgart is also a multi-functional envelope which plays with the perceptions of the observer. The 10 m high sculpture, again designed by Carsten Nicolai, consists of a very filigree inner structure of steel, which is encased by triangular glass panes. The panes have a special coating that creates a mirror effect during the day and transparency at night. This nocturnal transparency is achieved by the sources of light integrated in the base of the sculpture. A receiver in the interior of the sculpture reacts to electro-magnetic radiation from mobile telephones, laptops etc., and converts the radiation into light and audio signals. The sculpture and its environment are thus involved in an interaction.

Auch die Skulptur »polylit« auf dem Kleinen Schlossplatz in Stuttgart ist eine multifunktionale Hülle, die mit der Wahrnehmung des Betrachters spielt. Die 10 m hohe, ebenfalls von Carsten Nicolai entworfene Skulptur besteht aus einer sehr filigranen inneren Tragkonstruktion aus Stahl, welche mit dreieckigen Glasscheiben verkleidet ist. Diese Scheiben sind mit einer speziellen Beschichtung versehen, die am Tag für einen Spiegeleffekt und in der Nacht für Transparenz sorgt. Die nächtliche Transparenz wird durch Lichtquellen im Sockel erreicht. Im Inneren der Skulptur befindet sich ein Empfänger, der auf elektromagnetische Strahlung von Handys, Laptops etc. reagiert und sie in Licht- und Tonsignale umsetzt. Auf diese Weise treten die Skulptur und ihre Umgebung in Interaktion.

Materials

Very different materials have been used in the projects presented here. Depending on the architectural context, desired effect and structural requirements, these can be (to mention but a few) glass, wood, carbon fibers, woven fabric, steel, titanium or concrete, or a combination of several of these materials – ultimately decisive is that the materials are optimally aligned with the location and the specific construction task. The single-material concept is equally important. Composites of several materials, which can not be separated again and recycled, are not compatible with the concept of sustainability. For this reason, every material and every detail fulfills a pre-determined task. The component parts can therefore be produced exactly and simply combined. As a result, the structures are not only easy to mount and modify, they can also be dismantled and disposed of with very little effort.

In den hier vorgestellten Projekten kommen sehr unterschiedliche Materialien zum Einsatz. Je nach architektonischem Kontext, angestrebter Wirkung und tragwerksplanerischen Erfordernissen kann dies (um nur einige Beispiele zu nennen) Glas, Holz, Kohlefasern, Gewebe, Stahl, Titan oder Beton sein oder eine Kombination aus mehreren dieser Materialien – entscheidend ist letztlich, dass die Materialien für den Ort und die spezifische Konstruktionsaufgabe optimal abgestimmt sind. Wichtig ist ebenso der Ein-Werkstoff-Gedanke: Verbundwerkstoffe aus mehreren Materialien, die nicht mehr getrennt und wiederverwertet werden können, sind nicht mit dem Gedanken der Nachhaltigkeit vereinbar. Aus diesem Grund erfüllt jedes Material und jedes Detail eine bestimmte Aufgabe. Die Bauteile können deshalb präzise gefertigt und einfach gefügt werden. Die Strukturen sind dadurch nicht nur leicht zu montieren und zu verändern; sie können auch mit geringem Aufwand demontiert und entsorgt werden.

Gira
Architect: Ingenhoven Architekten, Düsseldorf, Germany
Planning time: 1999–2000
Construction time: 2000–2001
Tasks completed: full engineering service for structural engineering of loadbearing structure
Client: Giersiepen, Anlagen-, Verwaltungs- und Vertriebs GmbH & Co. KG, Radevormwald, Germany

Steel allows not only the design of wide-span, yet at the same time filigree, structures of the highest transparency. It also facilitates a high degree of prefabrication and recycling capability. For this reason, this material is a significant element in many of the projects presented here, whether as an element in the facade, in the structure or in the building shell.

One example that illustrates the great advantages of steel in terms of simple and economical, yet elegant structures is the production and office building of Gira in Radevormwald. The complex consists of two almost identical halls, which have a floor area of 71.50 × 22.50 m. The interior of the halls is completely pillar-free thanks to the characteristic steel arch construction of the supporting framework on the first floor. The existence of two structures – outer structure/inner structure – has several advantages. On the one hand, the interaction of the outer structure with the stage-like inner structure keeps it slender and elegant. This also enhances the effect of the facility in the landscape. On the other hand, the external structure also serves other functions besides its supporting function. It is simultaneously facade and roof, ventilation and water drainage. The inner structure is dedicated entirely to its function as a support structure for the production. A perfect symbiosis of technical purpose and aesthetic appearance.

Mit Stahl lassen sich nicht nur weit spannende und zugleich filigrane Strukturen höchster Transparenz konstruieren. Stahl erlaubt auch einen hohen Grad an Vorfertigung und Wiederverwertbarkeit. Deshalb ist dieser Werkstoff wesentliches Element in vielen der hier vorgestellten Bauten, sei es in der Fassade, im Tragwerk oder in der Gebäudehülle.

Anschauliches Beispiel für die großen Vorteile von Stahl in Bezug auf einfache und wirtschaftliche, aber dennoch elegante Strukturen ist das Produktions- und Bürogebäude von Gira in Radevormwald. Der Komplex besteht aus zwei annähernd baugleichen Hallen, die im Grundriss 71,50 × 22,50 m messen. Der Innenraum der Hallen ist aufgrund der charakteristischen Stahlbogenkonstruktion des Tragwerks im ersten Geschoss völlig stützenfrei. Das Vorhandensein von zwei Tragwerken – äußere Struktur/innere Struktur – hat mehrere Vorzüge: Zum einen ist die äußere Struktur im Zusammenwirken mit der ständerartigen inneren Struktur schmal und elegant gehalten. Zum anderen nimmt die äußere Struktur neben der tragenden auch noch weitere Funktionen wahr. Sie ist gleichzeitig Fassade und Dach, Lüftung und Wasserabführung. Die innere Struktur ist ganz auf die Funktion als Tragwerk für die Produktion konzentriert. Technische Bestimmung und ästhetische Anmutung sind hier eine gelungene Symbiose eingegangen.

Cologne-Bonn Car Park
Architect: Murphy/Jahn, Chicago, USA
Planning time: 1997–1998
Construction time: 1998–2000
Tasks completed: full engineering service except site observation for structural engineering of steel structures
Client: Flughafen Köln-Bonn GmbH, Germany

Steel also played a central role in the extension of the Cologne-Bonn airport. Werner Sobek was not only responsible for the transparent facades of Terminals 2 and M, but also designed the structure of the three new multi-story car parks built in the course of the extension. The multi-story car parks are equipped with an outer skin of high-grade steel mesh. While allowing free air circulation, this skin still protects the users against the elements. An easily installed and maintained outer skin which puts industrial filter fabrics of steel to a new use!

Auch bei der Erweiterung des Flughafens Köln-Bonn spielte Stahl eine zentrale Rolle. Werner Sobek zeichnete nicht nur für die transparenten Fassaden von Terminal 2 und M verantwortlich, sondern betreute auch Tragwerksentwurf und -planung der drei neuen Parkhäuser, die im Zuge der Erweiterung entstanden. Die Parkhäuser sind mit einer Außenhaut aus Edelstahlgewebe versehen, das eine freie Luftzirkulation zulässt, aber die Nutzer dennoch vor Wind und Regen schützt. Eine leicht zu installierende und zu wartende Außenhaut, die industrielle Filtergewebe aus Stahl für einen neuen Zweck nutzt.

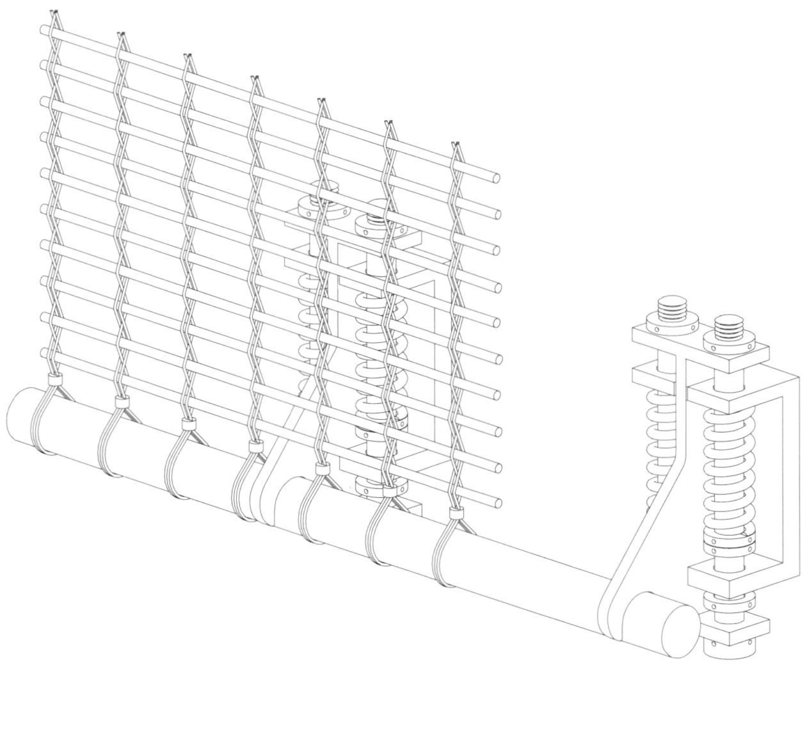

SWM Heilbronn
Architect: Dominik Dreiner, Gaggenau, Germany
Planning time: 2002 – 2003
Construction time: 2003 – 2004
Tasks completed: full engineering service except site observation for structural engineering of loadbearing structure
Client: Verband der Metall- und Elektroindustrie Baden-Württemberg, Germany

A steel screen was also used for the head office of Südwestmetall in Heilbronn. The structure of the building basically consists of a reinforced concrete construction. In the area of the floor-to-ceiling glazing on the longitudinal sides of the building, the ceiling rests on minimized stays of stainless steel. In order to underline the corporate identity of the employer association of the metalworking industry, the entire wall and roof surface of the reinforced concrete construction was clad with a woven fabric of filigree stainless steel strips.

Bei der Geschäftsstelle des Arbeitsgeberverbandes Südwestmetall in Heilbronn wurde ebenfalls ein Metallgewebe verwendet. Die Struktur des Gebäudes besteht im Wesentlichen aus einer Stahlbetonkonstruktion. Im Bereich der geschosshohen Verglasung an den Gebäudelängsseiten liegt die Decke auf minimierten Stützen aus Edelstahl auf. Um die Corporate Identity des Arbeitgeberverbandes der Metallindustrie zu unterstreichen, wurde die gesamte Wand- und Dachfläche der Stahlbetonkonstruktion mit einem Gewebe aus filigranen Edelstahlbändern überzogen.

Kunstmuseum Stuttgart
Architect: Hascher, Jehle & Assoziierte,
Berlin, Germany
Planning time: 1999–2002
Construction time: 2001–2004
Tasks completed: full engineering service except site observation for structural engineering of load-bearing structure, special structures, and facades
Client: Hochbauamt der Stadt Stuttgart, Germany

For many years, engineers totally neglected glass as a building material. In the past few years, however, a broad-based specialized knowledge has been acquired. Great progress has been made, in particular regarding the mechanical characteristics of the building material, the force transmission in glass component parts, and the development of a dimensioning process suited to the material. The result are light structures with a very high degree of dematerialization that would not have been feasible even a few years ago.

This extensive dematerialization is vividly illustrated by the new art museum in Stuttgart. Situated in the heart of the city, the building consists of an interior cube encased with Jurassic limestone, which is surrounded by a second skin of glass. The exterior facade reflects the environment during daytime; at night, it becomes almost invisible as a result of the interior lighting – a managed interplay between transparency and reflectivity, similar to that achieved in the »polylit« sculpture presented above. For visitors to the museum itself, the glass facade offers fascinating views of the surrounding city and a bright, light-flooded interior.

Nachdem Glas als Baustoff seitens der Ingenieure über viele Jahre hin völlig vernachlässigt worden war, konnte in den letzten Jahren ein breiter Kenntnisstand erarbeitet werden. Insbesondere bei den Fragen zum mechanischen Verhalten des Baustoffes, der Krafteinleitung in Glasbauteile und der Entwicklung eines dem Werkstoff angepassten Bemessungsverfahrens wurden große Fortschritte erzielt. Das Ergebnis sind lichte Strukturen mit einem sehr hohen Grad der Entmaterialisierung, wie er noch vor wenigen Jahren nicht realisierbar gewesen wäre.

Sehr anschaulich gemacht wird diese weitreichende Entmaterialisierung durch den Neubau des Kunstmuseums der Stadt Stuttgart. Im Herzen der Stadt erhebt sich dieser mit Jurakalk verkleidete Kubus, der von einer zweiten Haut aus Glas umgeben ist. Während die Glasfassade für außenstehende Betrachter tagsüber die Umgebung widerspiegelt, wird sie nachts durch die innenliegende Beleuchtung nahezu unsichtbar, scheint praktisch nicht mehr existent – ein gelungenes Wechselspiel zwischen Transparenz und Reflexivität, wie es auch bei der bereits beschriebenen Skulptur »polylit« zur Geltung kommt. Für den Besucher des Museums selbst bietet die Glasfassade faszinierende Ausblicke auf die umgebende Stadt und einen hellen, lichtdurchfluteten Raum.

Kunstmuseum Stuttgart

Mobile Life Campus
Architect: Henn Architekten, Berlin, Germany
Planning time: 2002–2004
Construction time: 2004–2005
Tasks completed: full engineering service except site observation for structural engineering of loadbearing structure and facades
Client: Wolfsburg AG, Germany

The Mobile Life Campus in Wolfsburg is another example of the comprehensive dematerialization of the building envelope made possible by the use of glass. The basic pattern of the buildings resembles a folded strip. The pattern is reflected in the facade, which is designed as a solid strip in interplay with closed and open surfaces. The interior is dominated by a glass-spanned atrium. The latter is both a general meeting place and a space for informal communication. The high quality of the interior space created by this high level of transparency is much appreciated by employees and visitors alike.

Ein anderes Beispiel für die weitreichende Entstofflichung der Gebäudehülle durch den Einsatz von Glas ist der Mobile Life Campus in Wolfsburg. Die Grundstruktur der Gebäude gleicht einem gefalteten Band und spiegelt sich in der Fassade wider, die als massives Band im Wechselspiel mit geschlossenen und geöffneten Flächen gestaltet ist. Das Innere wird geprägt durch ein glasüberspanntes Atrium, das sowohl Hauptversammlungsort als auch Raum für informelle Kommunikation ist. Die Transparenz führt zu einer hohen Innenraumqualität, die von allen Mitarbeitern und Besuchern außerordentlich geschätzt wird.

Papal Tent
Architect: Werner Sobek
Planning time: 2006
Construction time: 2006
Tasks completed: design, object planning and full engineering service
Client: Erzbischöfliches Ordinariat, Munich, Germany

Besides steel and glass, textiles are the most important materials in the work of Werner Sobek. It is the lightness of the material, as well as the numerous possibilities for shaping it, that are particularly fascinating. It was these qualities that have inspired designs such as the tent-roof for Pope Benedict XVI. On the occasion of his first visit to Germany in September 2006, the Pope celebrated a church service in front of 250,000 pilgrims. The cathedra of the Pope formed the highest point of the polygonal altar island and was spanned by a filigree membrane structure similar to the traditional Catholic canopy. The membrane structure had the form of a five point sail whose front-edge supporting stays raised up the major part of the expanse. The rear sail formation pulled the membrane surface down behind the altar island. The result was a wonderfully light superstructure, the supports of which were scarcely perceptible – but that nevertheless fulfilled the highest safety requirements.

Textilien sind neben Stahl und Glas die wichtigsten Materialien im Werk von Werner Sobek. Die Leichtigkeit ebenso wie die zahlreichen Möglichkeiten bei der Formgebung faszinieren an diesem Material besonders und inspirieren zu Entwürfen wie dem Zeltdach für Benedikt XVI. Anlässlich seines ersten Deutschlandbesuchs zelebrierte der Papst im September 2006 vor 250.000 Pilgern einen Festgottesdienst auf dem Freigelände der Neuen Messe München. Die Kathedra des Papstes bildete den höchsten Punkt der polygonalen Altarinsel und wurde von einer filigranen Membranstruktur überspannt, die dem traditionellen katholischen Traghimmel nachempfunden war. Die Membranstruktur hatte die Form eines Fünfpunktsegels, dessen vordere Randstützen einen Großteil der Fläche weit nach oben hoben. Die hintere Absegelung zog die Membranfläche hinter der Altarinsel nach unten, sodass eine für die Membrane optimale Geometrie entstand. Das Ergebnis war ein wunderbar leichter Überbau mit kaum mehr wahrnehmbarem Tragwerk, der dennoch höchsten Sicherheitsanforderungen gerecht wurde.

Papal Tent

Rotating Umbrellas
Architect: Werner Sobek
Planning time: 1997–2003
Construction time: 2003
Tasks completed: design, object planning and full engineering service

The dematerialization of membrane structures can be taken even further. A decisive factor in this context is the conceptual step towards building with energy, i.e. constructions that need virtually no permanent supporting structure and that owe their stability only to the energy induced according to specific requirements. The rib-free umbrella, developed jointly with Matthias Schuler, is an example of this: An object whose textile skin is spanned open through rotation and the centrifugal forces this creates. The mast of the umbrella is fixed, but the electrical motor at the top makes the textile surface (which is slightly reinforced at the hub) rotate. The textile surface is not even, but was calculated using a specially developed program based on the ratio of span/material weight/rotational speed. The calculations resulted in extremely complex, pagoda-shaped cross sections with an optimal stress distribution in the cloth. A convex edge was added to improve the aerodynamic stability. The result is a wonderfully light, floating cloud of material which captivates the observer: poetry in motion!

Der Gedanke der Entmaterialisierung von Membranstrukturen kann freilich noch weiter gedacht werden. Entscheidend ist hierbei der konzeptionelle Schritt hin zu einem Bauen mit Energie, das praktisch ganz ohne stützende Struktur auskommt und seine Stabilität einzig und allein der – je nach Bedarf induzierten – Bewegungsenergie eines Objekts verdankt. Beispiel hierfür ist der spantenlose Schirm: ein zusammen mit Matthias Schuler entwickeltes Objekt, dessen textile Haut durch Rotation und die dabei entstehenden Zentrifugalkräfte aufgespannt wird. Der Mast des Schirms ist starr, aber in seinem Kopf befindet sich ein Elektromotor, der die im Nabenbereich leicht verstärkte textile Fläche in Rotation versetzt. Diese textile Fläche ist nicht eben, sondern wurde mit speziell entwickelten Computerprogrammen aus dem Verhältnis Spannweite/Stoffgewicht/Drehzahl berechnet. Dadurch ergeben sich äußerst komplexe, pagodenförmige Querschnitte mit einer optimalen Spannungsverteilung im Tuch. Zur Verbesserung der aerodynamischen Stabilität wurde darüber hinaus eine gewölbte Umrandung der Struktur entwickelt. So entstand eine leichte, schwebende Wolke aus Stoff, die den Betrachter allein durch ihre poetische Kraft gefangen nimmt.

Grenzen (er)leben
Architect: Atelier Brückner, Stuttgart, Germany
Planning time: 2001–2002
Construction time: 2002
Tasks completed: full engineering service for structural engineering of loadbearing structure and facades
Client: Interkantonale Trägerschaft »Grenzen (er)leben«, Switzerland

Wood can be used in the most diverse array of forms and functions: as a roof bearing structure (as in the case of the New Trade Fair in Hamburg), as a facade element (as in the case of the Burda multi-story car park in Offenburg) or even as part of an installation like the pavilion »Experience (live) limits« developed by Uwe Brückner. The two-story exhibition pavilion, which was part of the Swiss National Exhibition 2002, was located on a platform in Lake Biel. It consisted of an art room resting on 84 timber props, each 4 m in length. The idea was for visitors to the exhibition to experience limits in a new way, including the limits of rooms themselves. This intention should also be reflected in the structure. The pole forest was, therefore, a consciously fluid boundary between internal and external space, between sky and water, between solid and movable. The pole forest continued in front of the actual pavilion as a sculptural installation.

Holz kann in den verschiedensten Formen und Funktionen zum Einsatz kommen: als Dachtragwerk (wie bei der Neuen Messe in Hamburg), als Fassadenelement (wie beim Burda-Parkhaus in Offenburg) oder auch als Teil einer Installation – wie z.B. beim vom Atelier Brückner entworfenen Pavillon »Grenzen (er)leben«. Der zweigeschossige Ausstellungspavillon, der Teil der Schweizer Landesausstellung 2002 war, befand sich auf einer Plattform im Bieler See. Er bestand aus einem Kunstraum, der auf 84 Holzstützen von jeweils 4 m Länge, dem so genannten Stangenwald, auflag. Die Besucher der Ausstellung sollten Grenzen, auch die von Räumen, neu (er)leben; diese Intention sollte auch im Tragwerk ablesbar sein. Der Stangenwald war deshalb eine bewusst fließende Grenze zwischen innen und außen, zwischen Himmel und Wasser, zwischen fest und beweglich. Vor dem eigentlichen Pavillon setzte er sich als skulpturale Installation fort.

Cruise Center
Architect: Renner, Hainke, Wirth, Hamburg, Germany
Planning time: 2003–2004
Construction time: 2004
Tasks completed: full engineering service for structural engineering of loadbearing structure and roof
Client: HafenCity Hamburg GmbH, Germany

Wood also played an important role in another temporary structure, the cruise ship terminal in Hamburg's redeveloped port area. The terminal will be used until 2010 by cruise ship passengers for check-in and baggage registration. The hall's floor plan measures approx. 23 × 67 m; the height of the construction is 9 m. In order to guarantee fast and economical construction, the walls of the cruise ship terminal were made from standard steel containers, stacked in three layers on top of each other. The roof structure consists of wooden framework ties with an overhang on both sides and a covering of panels (upper surface) and of double linking plates (underside), which are illuminated from the inside at night. This lightweight and efficient roof structure was pre-assembled in parts and then mounted with cranes within hours.

Auch in einem anderen temporären Bau, dem Kreuzfahrtterminal in der neuen HafenCity Hamburg, spielt Holz eine wichtige Rolle. Das Terminal dient bis zum Jahr 2010 als Anlaufstelle für Kreuzfahrtpassagiere zur Registrierung und Gepäckaufgabe. Die Außenabmessungen der Halle betragen im Grundriss ca. 23 × 67 m, die Höhe der Konstruktion liegt bei 9 m. Um einen schnellen und preiswerten Aufbau zu gewährleisten, wurden die Wände des Kreuzfahrtterminals aus handelsüblichen, in drei Lagen übereinander gestapelten Stahlcontainern geplant. Das Dachtragwerk besteht aus Holzfachwerkbindern mit beidseitigem Dachüberstand und einer Eindeckung aus Paneelen (oberseitig) bzw. aus Doppelstegplatten (unterseitig), die nachts von innen beleuchtet werden. Diese leichte und effiziente Dachstruktur wurde in Teilen vormontiert und dann mit Kränen innerhalb weniger Stunden fertig montiert.

Hamburg
Cruise Center

Mercedes-Benz Museum
Architect: UN studio van Berkel & Bos, Amsterdam, Netherlands
Planning time: 2002–2004
Construction time: 2004–2006
Tasks completed: full engineering service for structural engineering of loadbearing structure; full engineering service except site observation and tender documents for facades
Client: DaimlerChrysler Immobilien GmbH, Berlin, Germany

Many observers are surprised again and again just how light and elegant the material concrete can appear. Spatially-curved surfaces in exposed concrete quality are one example; wide-spanning shell support structures another. In order to achieve such results, however, architect and engineer must already work hand in hand in the initial design phases. The new Mercedes-Benz museum in Stuttgart impressively demonstrates the high structural quality that can be achieved. The highly-complex building consists of a wide-spanning reinforced concrete structure on 8 levels. The individual levels span widths of up to 30 m without support. The floor layout, which resembles a three-leafed clover, has an edge length of approx. 80 m on each side. The height of the building is approx. 47 m. All facade supports are slanted and accommodate the »folded edges« of the glass facade wrapped around the building. This results in a different length, as well as different setting angles, for each of the pillars. The structure of the building, reminiscent of a double helix, is reflected not only in the facade with its transparent and closed surfaces. It is also an integral part of the functional concept. Divided into collections of legendary cars oriented toward the inside and chronological collections opened toward the outside, there are two thematic walkways through the museum that lead the visitor through the entire double helix. Both walkways can be interrupted on every level and continued in the other part of the museum. Due to its complex geometry, the entire building was planned in 3D.

Überraschend für jeden Betrachter ist immer wieder, wie leicht und elegant der kompakte Werkstoff Beton wirken kann. Räumlich gekrümmte Flächen in Sichtbetonqualität finden sich hierbei ebenso wieder wie weitspannende Schalentragwerke verschwindender Dicke.

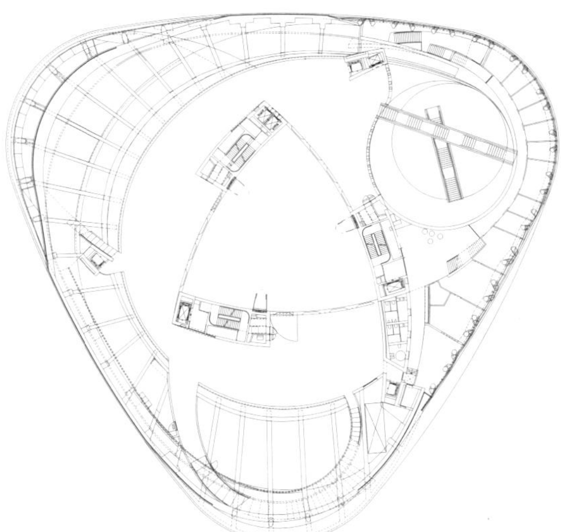

Mercedes-Benz Museum

Das neue Mercedes-Benz Museum in Stuttgart demonstriert eindrücklich die hohe Gestaltqualität, die durch eine enge Zusammenarbeit von Architekt und Ingenieur bereits in der ersten Entwurfsphase erreicht werden kann. Das hochkomplexe Gebäude besteht aus einer weitspannenden Stahlbetonkonstruktion mit zum Teil räumlichen Stahlbetontragwerken auf acht Ebenen, die ab der Ebene 2 an einer Doppelhelix angelenkt sind. Die einzelnen Ebenen spannen bis zu 30 m stützenfrei. Der an ein dreiblättriges Kleeblatt erinnernde Grundriss hat eine Kantenlänge von jeweils ca. 80 m. Die Höhe des Gebäudes beträgt ca. 47 m.

Alle Fassadenstützen sind im Raum geneigt und nehmen die »Faltungskanten« der sich um das Gebäude windenden Glasfassade auf. So ergeben sich für alle Stützen neben unterschiedlichen Anstellwinkeln auch unterschiedliche Stützenlängen. Die einer Doppelhelix nachempfundene Struktur des Gebäudes spiegelt sich nicht nur in der Fassade mit ihren transparenten bzw. geschlossenen Flächen wider, sondern ist auch integraler Teil des funktionalen Konzeptes. Aufgeteilt in nach innen orientierte Mythosflächen und nach außen geöffnete Sammlungen gibt es zwei thematische Durchgänge durch das Museum, die den Besucher durch die gesamte Doppelhelix führen. Beide Durchgänge können auf jeder Ebene unterbrochen und im jeweils anderen Teil des Museums fortgesetzt werden. Aufgrund seiner komplexen Geometrie wurde das gesamte Gebäude in 3-D geplant.

Mercedes-Benz Museum

P & C Lübeck
Architect: Ingenhoven Architekten, Düsseldorf, Germany
Planning time: 2000–2003
Construction time: 2003–2005
Tasks completed: schematic design and design development for structural engineering of loadbearing structure
Client: Marktplatz Lübeck GbR, Germany

While the complex structure of the Mercedes-Benz museum is revealed to the observer only after entering the building, the filigree concrete shell structure over the Peek & Cloppenburg department store in Lübeck is already visible from a distance. In direct proximity to the UNESCO protected buildings of the old town, a new business building has been carefully placed in the historically important city center. Architect Christoph Ingenhoven attached great importance to the roof. It consists of 16 reinforced concrete shells, each just 14 cm thick. The shells have a trapezoidal layout and are arranged in alternate directions. The spans of the individual shells are 8.75 m in transverse direction and up to 28 m in longitudinal direction. The apparently light, harmoniously curved roof thus sets a succinct architectural accent and yet it is perfectly integrated into the surrounding cityscape.

Während sich die komplexe Struktur des Mercedes-Benz Museums dem Betrachter erst nach Betreten des Bauwerks erschließt, ist die filigrane Betonschalenstruktur über dem Peek & Cloppenburg Kaufhaus in Lübeck schon von Weitem sichtbar. In unmittelbarer Nähe zu den historisch bedeutsamen Gebäuden von Lübecks Altstadt entstand nach den Entwürfen von Christoph Ingenhoven ein neues Geschäftshaus, das sich behutsam in die unter UNESCO-Schutz stehende Innenstadt einzufügen hatte. Eine große Bedeutung kam daher dem Dach zu. Es besteht aus 16 Stahlbetonschalen mit einer Dicke von nur 14 cm, die einen trapezförmigen Grundriss besitzen und in alternierender Richtung angeordnet sind. Die Spannweiten der einzelnen Schalen betragen in Querrichtung 8,75 m und in Längsrichtung bis zu 28 m. Das leicht anmutende, harmonisch geschwungene Dach setzt so einen prägnanten architektonischen Akzent, fügt sich aber dennoch hervorragend in die umgebende Stadtlandschaft ein.

Mae West
Artist: Rita McBride
Planning time: 2005–2006
Construction time: 2008
Tasks completed: object planning and full engineering service
Client: Landeshauptstadt München, Germany

Concrete, glass, steel, wood, and textiles are only some of the numerous materials employed in projects by Werner Sobek. To gain an adequate understanding of his work it is also important to take a look at some of the innovative materials that are used (often for the first time in the construction industry). This includes glass-fiber reinforced plastics, as well as carbon-reinforced glass panes or epoxy sandwich panels.
An example of this innovative employment of materials is the sculpture »Mae West« in Munich. This sculpture, designed by the American artist Rita McBride, is an important part of an urban regeneration project in the city of Munich. It is entirely made of carbon fiber tubes. The work of art has the form of a hyperbolic paraboloid and consists of a double support-structure with a total height of 52 m. The diameter of the sculpture is 32 m at the base and 19.50 m at the top – after the diameter has tapered to only about 7.50 m at approx. two-thirds of the total height. As a result of the filigree lattice structure, the sculpture appears permeable and open, despite its impressive height.

Beton, Glas, Stahl, Holz und Textilien sind nur einige der zahlreichen Materialien, die in den Projekten von Werner Sobek verwendet werden. Wichtig für ein angemessenes Verständnis seines Œuvres ist auch ein Blick auf die Werkstoffe, die erstmals im größeren Umfang im Bauwesen eingesetzt worden sind. Glasfaserverstärkte Kunststoffe gehören ebenso dazu wie karbonbewehrte Glasscheiben oder Epoxysandwichplatten.
Ein Beispiel für diesen innovativen Einsatz von Materialien ist die Skulptur »Mae West« in München: Die hier verwendeten Materialien haben bisher kaum Einzug ins Bauwesen gehalten. Die von der amerikanischen Künstlerin Rita McBride entworfene Skulptur aus Karbonfaserrohren wird im Rahmen einer städtebaulichen Neugestaltung in der Münchner Innenstadt aufgestellt werden. Das Kunstwerk hat die Form eines hyperbolischen Paraboloids und besteht aus einer doppelten Tragstruktur mit einer Gesamthöhe von 52 m. Der Durchmesser der Skulptur beträgt 32 m am Fußpunkt und 19,50 m an der Spitze – nachdem sich der Durchmesser auf ca. zwei Drittel der Gesamthöhe auf nur rund 7,50 m verjüngt hat. Durch die filigrane, aufgelöste Gitterstruktur erscheint die Skulptur trotz ihrer Höhe durchlässig und offen.

Design

Through the planning of structure and facade, Werner Sobek plays an important role in shaping the form and function of numerous buildings. However, the studio also delivers complete designs for products such as chairs or cable end fittings, just as it does for bridges, towers, trade fair stands, temporary structures, houses or urban objects. A common feature in all these designs is their high structuring quality, the shaping, right down to the smallest detail, and the high level of functionality. Everything that is superfluous is omitted. Where possible, work is carried out with standardized and pre-assembled elements, which guarantee high levels of cost-effectiveness and high quality. Form, function and structure thus converge to form a perfect unit.

Durch die Planung von Tragwerk und Fassade nimmt Werner Sobek wesentlichen Einfluss auf Form und Funktion zahlreicher Objekte. Aber auch die ganzheitliche Gestaltung von Produkten wie Stühlen oder Seilendbeschlägen gehört zum Leistungsspektrum, ebenso wie der Entwurf und die architektonische Planung von Brücken, Türmen, Messeständen, temporären Bauten, Häusern oder städtebaulichen Objekten. Gemeinsam ist allen diesen Entwürfen ihre hohe Gestaltqualität, die Durchformung bis ins kleinste Detail und ihre ausgeprägte Funktionalität. Alles Überflüssige ist weggelassen. Wo immer möglich wird mit standardisierten und vorgefertigten Elementen gearbeitet, die für große Wirtschaftlichkeit und hohe Qualität sorgen. Form, Funktion und Struktur verbinden sich so zu einer perfekten Einheit.

Mercedes-Benz Wings
Architect: Werner Sobek
Planning time: 2002–2003
Construction time: 2003–2005
Tasks completed: design, object planning and full engineering service
Client: DaimlerChrysler AG, Stuttgart, Germany

The extremely high degree of quality and cost-effectiveness which becomes possible through system formation and prefabrication is demonstrated particularly effectively in the wing elements that were developed as a new standard element for all trade fair stands of Mercedes-Benz worldwide. By always using the same, characteristic basic elements in every trade fair stand, brand recognition is ensured and an unmistakable and uniform appearance guaranteed. At the same time, each new trade fair stand can easily be individualized, simply by varying some of the basic elements such as wing geometry, surface structure and materials.

Der extrem hohe Grad an Qualität und Wirtschaftlichkeit, der durch Systembildung und Vorfertigung möglich wird, lässt sich besonders gut an den Wing-Elementen ablesen, die als neue Standardkomponenten für alle Messeauftritte von Mercedes-Benz weltweit entwickelt wurden. Jeder Messestand erhält durch die Verwendung immer gleicher, charakteristischer Grundelemente einen hohen Wiedererkennungswert und gewährleistet ein unverwechselbares und einheitliches Erscheinungsbild. Gleichzeitig können die einzelnen Messeauftritte durch Variationen der Grundelemente in Hinblick auf Geometrie, Materialität und Oberflächenausbildung äußerst flexibel und immer wieder neu – aber dennoch wirtschaftlich – gestaltet werden. Möglich ist dies durch ein Tragsystem, das mit den unterschiedlichsten Oberflächen belegt werden kann.

cia
CLS

Cable Fittings

Architect: Werner Sobek
Planning time: 2000–2001
Construction time: 2001
Tasks completed: design, object planning and full engineering service
Client: Brugg Drahtseile AG, Birr, Switzerland

The cable fittings designed for a Swiss cable manufacturer unite cost-effectiveness with high structural quality and minimalism. Following the principles of lightweight construction, the design was optimized in a multi-stage process. Stress concentrations were thus kept to an absolute minimum and less loaded areas could be eliminated. In addition, the system can be structured flexibly through the exchangeability of individual parts. Unique to this cable-end fitting, it is also possible to pre-stress the cable using a specially developed device.

Auch der für einen Schweizer Seilhersteller gestaltete Seilendbeschlag vereint Wirtschaftlichkeit mit hoher Gestaltqualität und einer Reduktion auf das Wesentliche. Gemäß den Prinzipien des Leichtbaus wurde das Design in einem mehrstufigen Prozess optimiert: Spannungskonzentrationen wurden somit auf ein absolutes Minimum zurückgeführt, wenig belastete Bereiche konnten ausgespart werden. Außerdem wurde das System durch die Austauschbarkeit einzelner Teile flexibel gestaltet. Einmalig an diesem Seilendbeschlag ist darüber hinaus die Möglichkeit, durch eine speziell entwickelte Einrichtung die Vorspannung auf das Seil über den Seilendbeschlag selbst aufzubringen.

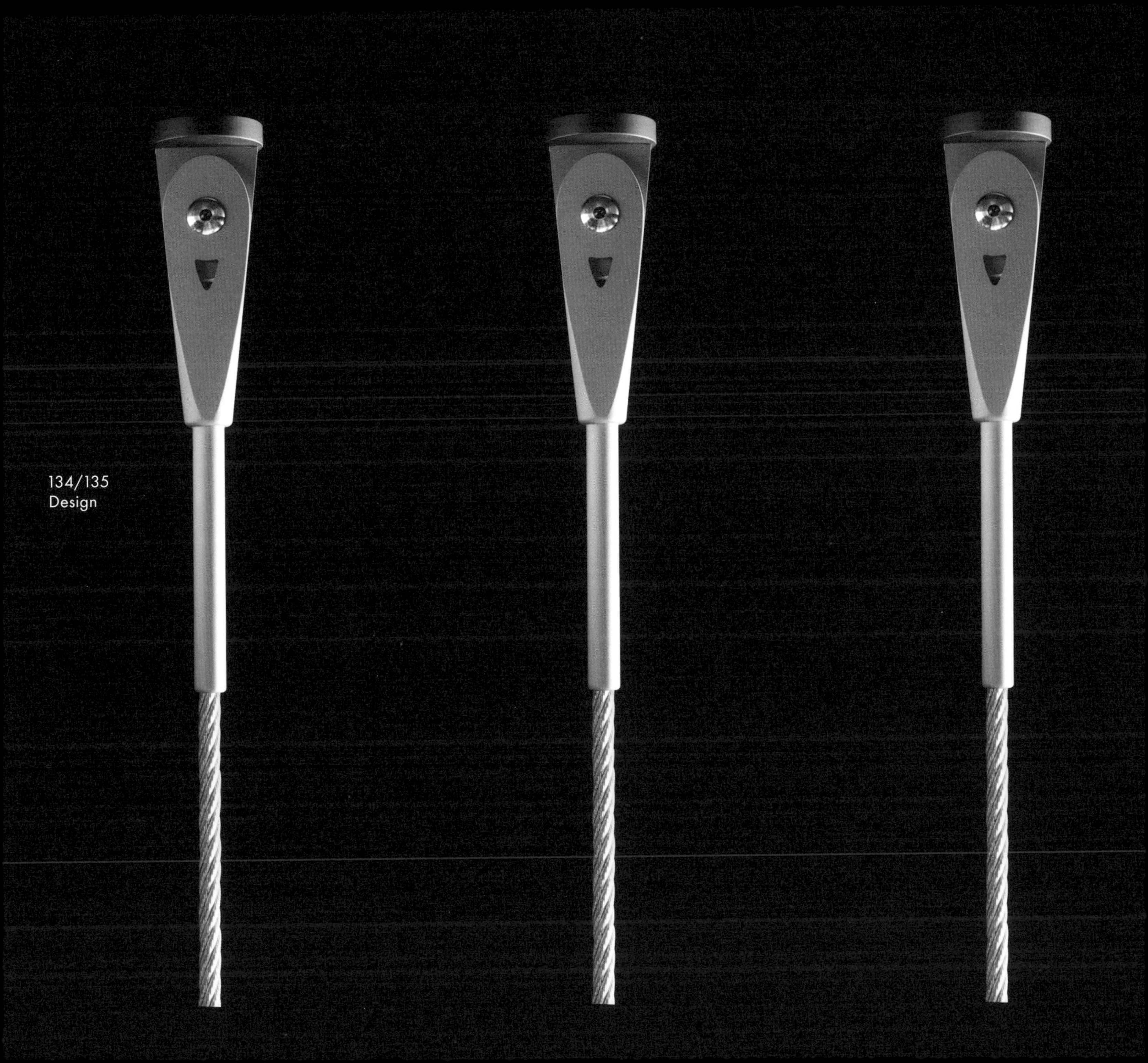

Airport Chair
Architect: Werner Sobek
Planning time: 2000–2001
Construction time: 2001–2002
Tasks completed: design, object planning and full engineering service

A minimizing of the constructive cross sections in accordance with the usual principles of lightweight construction can also be found in the Airport Chair. Radiating the lightness of a curved strip, its form, materialism and jointing consciously draw analogies to aircraft construction. The elegant form is constructed using rolled aluminum sheet metal. The joints are surface-flush and precise. The result is a wonderfully light place to rest and relax in, perfectly adapted to the requirements of the user.

Eine Minimierung der konstruktiven Querschnitte gemäß den im Leichtbau üblichen Prinzipien zeichnet auch den Airport Chair aus. Er strahlt die Leichtigkeit eines im Raum geschwungenen Bandes aus und weist in Form, Materialität und Fügung bewusst gewählte Analogien zum Flugzeugbau auf. Die elegante Form wird durch gewalzte Aluminiumbleche gebildet. Die Fügungen sind flächenbündig und präzise. Das Ergebnis ist ein leichter, an die Bedürfnisse des Nutzers angepasster Ort zum Verweilen und Entspannen.

Mercedes-Benz Motorsport
Architect: Werner Sobek
Planning time: 2004 – 2005
Construction time: 2005
Tasks completed: design, object planning and full engineering service
Client: Mercedes-Benz, Stuttgart, Germany

Trade fair stands such as »MB Motorsport Set« or »Nautilus« are characterized not only by their painstaking attention to detail or by their innovative surfaces and structural designs. As a result of their high degree of prefabrication and sophisticated assembly concepts, they can also be assembled and dismantled quickly, exactly and economically. An example of this is the Motorsport Set, which is used for the brand launch of Mercedes-Benz at international racing events, such as Formula 1 and DTM Racing. The set consists of two building modules – the Mbox and the Unit – and the wings. The modular-designed wings structure the interior space and can be used as a comprehensive element for brand communication, ensuring high brand recognition. For every racing series, a Motorsport Set with different exhibition areas is used. In the process, the buildings are assembled and dismantled again approx. fifty times and transported over thousands of kilometers.

Messestände wie das »MB Motorsport Set« oder der »Nautilus« zeichnen sich nicht nur durch höchste Detailqualität und innovative Oberflächen und Formelemente aus. Sie sind durch ihren hohen Grad der Vorfertigung und die ausgeklügelten Montagekonzepte auch schnell, präzise und wirtschaftlich auf- und abzubauen. Beispiel hierfür ist das Motorsport Set, das dem Markenauftritt von Mercedes-Benz bei internationalen Rennveranstaltungen wie Formel 1 und DTM-Rennen dient. Das Set besteht aus zwei Gebäudemodulen – der Mbox und der Unit – und den so genannten Wings. Die modular konzipierten Wings fungieren als raumbildende Elemente, die den Innenraum strukturieren, und werden als übergreifendes Element zur Markenkommunikation genutzt. Somit schaffen sie einen hohen Wiedererkennungswert der Marke. Pro Rennserie wird je ein Motorsport Set mit variierenden Ausstellungsflächen betrieben. Hierbei werden die Gebäude ca. fünfzig Mal auf- und wieder abgebaut und legen dabei tausende Transportkilometer zurück.

Mercedes-Benz

Mero
Architect: Werner Sobek
Planning time: 2001 – 2002
Construction time: 2002
Tasks completed: design, object planning and full engineering service
Client: Mero Systeme GmbH & Co. KG, Würzburg, Germany

The trade fair stand »Nautilus« was developed for the company Mero. Thanks to the Mero rod-node system, the degree of standardization was extremely high. But the most distinguishing feature of the stand was surely the special surface created by a vacuum stabilized membrane. As a result, the trade fair stand developed an extremely high tactile quality inviting fairgoers to stay and touch. The supporting element of the stand was a spatially-curved shell of 28 m length and 5.20 m height. A special foil was placed over this structure. The foil was attached to the lower end of the structure and sealed. Then the air was sucked from the space between foil and structure. The wrinkles deliberately generated during this process provided the surface with a striking and poetic texture.

Der Messestand »Nautilus« wurde für die Firma Mero entwickelt. Herausragendes Kennzeichen war – neben dem hohen Maß an Standardisierung, das durch die Verwendung des Mero Stab-Knoten-Systems erreicht werden konnte – die besondere Oberfläche, die durch eine vakuumstabilisierte Membran entstand. Der Messestand entwickelte hierdurch eine äußerst große haptische Qualität, die die Messebesucher zum Verweilen und Berühren einlud. Tragendes Element des Standes war eine räumlich gekrümmte Schale von 28 m Länge und 5,20 m Höhe, über die eine spezielle Folie gelegt wurde. Diese Folie wurde am unteren Ende der Struktur befestigt und abgedichtet. Anschließend wurde die Luft aus dem Zwischenraum zwischen Folie und Struktur gesaugt. Die dabei gezielt erzeugten Falten sorgten für eine spezifische, poetisch anmutende Textur der Oberfläche.

Deutsche Post
Architect: Werner Sobek
Planning time: 1999–2000
Construction time: 2000–2001
Tasks completed: design, object planning and full engineering service
Client: Deutsche Post Immobilien GmbH, Bonn, Germany

Specific tactile qualities also played a role in developing the trade fair stand for Deutsche Post. On the stand a large number of wing-shaped wall units with black, rubber-like surfaces were configured to a space-confining spiral. Eight of the elements were equipped on one side with colored light-emitting diodes and thus formed a high-quality screen.
In addition, smaller screens, extendable from black columns, provided the visitors with information. The consulting areas were equipped with specially developed, curved furniture. The furniture was clad with a rubber-like coating characterized by its high tactile quality, as well as its high durability.

Spezifische haptische Qualitäten spielten auch bei dem Messestand eine Rolle, der für die Deutsche Post entwickelt wurde. Auf dem Stand wurde eine Vielzahl flügelförmiger Wandelemente mit schwarzen, gummiartigen Oberflächen zu einer raumbildenden Spirale konfiguriert. Acht der Elemente wurden einseitig mit farbigen Leuchtdioden bestückt und formten so einen hochwertigen Bildschirm. Zusätzlich informierten kleinere, aus schwarzen Säulen herausfahrbare Bildschirme die Besucher. Die Rückzugsbereiche für persönliche Beratung wurden mit eigens entworfenen, räumlich geschwungenen Möbeln ausgestattet, die ebenfalls einen speziellen gummiartigen Überzug erhielten, der sich durch seine hohe haptische Qualität sowie durch eine große Belastbarkeit auszeichnete.

R128
Architect: Werner Sobek
Planning time: 1998–1999
Construction time: 1999–2000
Tasks completed: design, object planning and full engineering service for loadbearing structure, special structures, and facades
Client: Ursula and Werner Sobek, Stuttgart, Germany

The work of Werner Sobek is characterized by structuring quality and sustainability, systematic and clarity, cost-effectiveness and precision, and is guided not least by the requirements of the user. Residential buildings allow an excellent synopsis of all these values and qualities. In these buildings, Werner Sobek's design, functionality, and consideration of ecological aspects are combined in perfect symbiosis.

Let's take R128, a completely glazed, four-story building at the edge of the Stuttgart basin, as a first example. Only a few years old, R128 has already become an icon of modern architecture. R128 is a completely recyclable, zero-energy house which produces no emissions. Thanks to its modular structure, most of it could be pre-assembled and then finished on the construction site within a very short space of time. Made exclusively of modular components, the house is not only easy to assemble and dismantle, but also completely recyclable.

Das Werk von Werner Sobek ist geprägt durch Gestaltungsqualität und Nachhaltigkeit, Systematik und Klarheit, Wirtschaftlichkeit und Präzision und – last, but not least – Orientierung an den Bedürfnissen der Nutzer. Eine sehr gute Zusammenschau auf all diese Werte und Qualitäten ermöglichen Wohnhäuser. In diesen Objekten kommen Design, Funktionalität und Berücksichtigung ökologischer Aspekte in perfekter Symbiose zusammen.

Als erstes Beispiel sei hier nur kurz an R128 erinnert, ein komplett verglastes viergeschossiges Gebäude am Rande des Stuttgarter Talkessels. Zwar erst einige Jahre alt, ist R128 bereits zu einer Ikone der modernen Architektur geworden. R128 ist ein vollkommen wiederverwertbares, im Betrieb emissionsfreies Nullenergiehaus. Durch seinen modularen Aufbau konnte es zu großen Teilen vorgefertigt und auf der Baustelle innerhalb kurzer Zeit fertiggestellt werden.

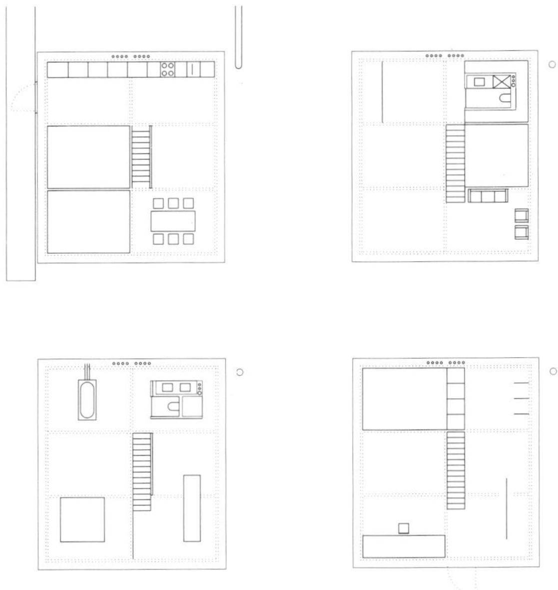

H16
Architect: Werner Sobek
Planning time: 2002 – 2005
Construction time: 2005 – 2006
Tasks completed: design, object planning and full engineering service for loadbearing structure, special structures, and facades
Client: H. & G. Link, Balingen, Germany

H16 is another emission-free, zero-energy house which offers the highest level of living comfort and great structuring quality. H16 consists of two contrasting cubes. A transparent all-glass cube provides a generous room with a flexible floor layout for eating and living. A black cube made from precast concrete elements contains the private rooms and guarantees intimacy and possibilities for withdrawal. A stone wall is used for the inclined close-off, thus creating an open space around the cubes. A specially designed air-conditioning concept enables emission-free heating and cooling.

Auch H16 ist ein emissionsfreies Nullenergiehaus, das höchsten Wohnkomfort und große Gestaltungsqualität aufweist. H16 besteht aus zwei kontrastierenden Kuben. Ein transparenter Ganzglas-Kubus bietet einen großzügigen Raum mit flexiblem Grundriss für Essen und Wohnen. Der schwarze Kubus aus Sichtbetonfertigteilen beinhaltet die privaten Räume und sorgt für Intimität und Rückzugsmöglichkeiten. Eine Natursteinmauer dient dem Hangabschluss und stellt den Raum für die Kuben frei. Das speziell abgestimmte Klimakonzept ermöglicht emissionsfreies Heizen und Kühlen.

H16

R129
Architect: Werner Sobek
Planning time: 2001-2010
Construction time: 2010-2012
Tasks completed: design, object planning and full engineering service

An even more radical vision for a life in nature, separated from the outside world only by means of a virtually imperceptible envelope, is R129. R129 is a lenticular body with a completely flexible floor layout. The transparent envelope is made of plastic, the mounting structure of carbon hollow support beams. In addition, the envelope can be darkened gradually or made completely opaque using a switchable electrochrome foil. The base slab, consisting of a carbon body, bears a heated floor as well as a technical floor. The latter offers storage facilities as well as connections for power, water, compressed air, and communication media. The interior area has no fixed partitions. The sanitary and kitchen installations are located in a moveable central unit. Once all the necessary preparatory work and research is completed, Werner Sobek will make this vision come true.

Eine noch radikalere Vision für ein Wohnen in der Natur, nur durch eine praktisch nicht mehr wahrnehmbare Hülle von der Außenwelt getrennt, ist R129, ein linsenförmiger Körper mit vollkommen flexiblem Grundriss. Die transparente Hülle besteht aus Kunststoff, das Traggerüst aus Karbonhohlträgern. Durch eine schaltbare elektrochrome Folie kann die Hülle darüber hinaus abschnittsweise oder in toto abgedunkelt oder komplett undurchsichtig geschaltet werden. Die aus einem Karbonkorpus bestehende Bodenplatte trägt einen Heizboden sowie einen Technikboden, der Speichermöglichkeiten und Anschlüsse für Strom, Wasser, Druckluft und Kommunikationsleitungen bietet. Der Innenbereich kennt keine festen Abtrennungen. In einer verschiebbaren, ortsunabhängigen Zentraleinheit befinden sich Sanitär- und Kücheninstallationen. Sobald alle nötigen Vorarbeiten und Recherchen abgeschlossen sind, wird Werner Sobek diese Vision Wirklichkeit werden lassen.

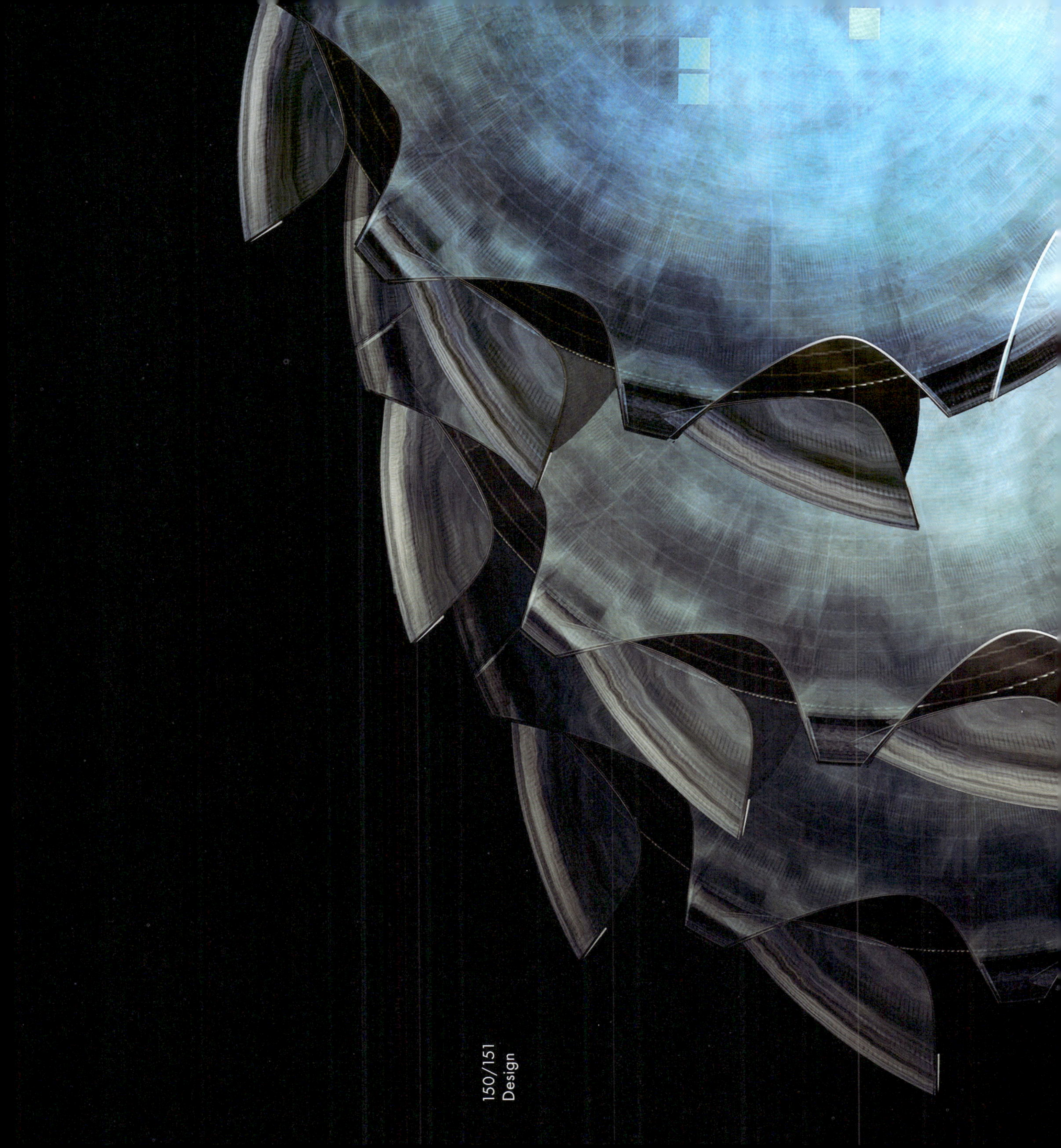

Werner Sobek stands throughout the world for engineering, design, and green technologies. The firm has offices in Stuttgart, Cairo, Dubai, Frankfurt, Khartoum, Moscow, and New York. The work of Werner Sobek is defined by premium design on the basis of high-class engineering combined with sophisticated green technologies. Founded in 1992, the studio works on all types of buildings and materials. Special emphasis lies on lightweight load-bearing structures, highrise buildings, and transparent facade systems. Another focus of the activities of Werner Sobek are special structures in steel, glass, titanium, textiles, and wood as well as advanced concepts for sustainable buildings.

Owner of the firm is Prof. Dr.Werner Sobek. He is architect, consulting engineer, proof engineer for all fields of structural engineering, and holds a Shanghai Engineering License as well as a Structural Engineering License for Illinois. As a successor to architect Frei Otto and structural engineer Jörg Schlaich, Werner Sobek heads the Institute for Lightweight Structures and Conceptual Design at the University of Stuttgart.

Werner Sobek steht weltweit für Engineering, Design und Green Technologies. Die Firmengruppe hat Niederlassungen in Stuttgart, Dubai, Frankfurt, Kairo, Khartoum, Moskau und New York. Die Arbeiten des Büros zeichnen sich durch hochklassige Gestaltung auf der Basis von herausragendem Engineering und ausgeklügelten Konzepten zur Minimierung von Energie- und Materialverbrauch aus. Das 1992 gegründete Unternehmen bearbeitet alle Typen von Bauwerken und Materialien. Besondere Schwerpunkte liegen im Hochhausbau, in der Fassadenplanung, in Sonderkonstruktionen in Stahl, Glas, Titan, Stoff und Holz sowie in der Planung von nachhaltigen Gebäuden.

Inhaber der Firmengruppe ist Prof. Dr. Werner Sobek. Er ist Architekt, beratender Ingenieur, Prüfingenieur für Baustatik aller Fachrichtungen und besitzt sowohl eine Shanghai Engineering License als auch eine Structural Engineering License für den US-Bundesstaat Illinois. Als Nachfolger von Frei Otto und Jörg Schlaich leitet Werner Sobek das Institut für Leichtbau Entwerfen und Konstruieren an der Universität Stuttgart.

Offices
Projects

Prof. Dr. Werner Sobek

1953 Born in Aalen/Germany. 1974–1980 Studied structural engineering and architecture at the University of Stuttgart, Germany. 1980–1986 Post-graduate fellow in research project »Wide-Span Lightweight Structures« at the University of Stuttgart, Germany. 1987 PhD in structural engineering. 1987–1991 Structural engineer in the office of Schlaich, Bergermann & Partner, Stuttgart, Germany. 1991–1994 Professor at the University of Hanover, Germany (successor to Bernd Tokarz) and director of the Institute for Structural Design and Building Methods. 1992 Founded Werner Sobek Stuttgart. 1994 Professor at the University of Stuttgart, Germany (successor to Frei Otto), Director of the Institute for Lightweight Structures. 1998–2006 Member of the board of the Chamber of Engineers of Baden-Württemberg. 1998 Proof engineer for all types of structures and materials. 1999 Founded Werner Sobek Design; exhibition »Archineering«, Municipal Museum Leverkusen, Germany. 2000 Second chair at the University of Stuttgart: successor to Jörg Schlaich. Foundation of the Institute for Lightweight Structures and Conceptual Design; Shanghai Engineering License. 2000–2001 Visiting professor at the University of Graz, Austria. 2001 Founded Werner Sobek Frankfurt; Structural Engineering License for the State of Illinois. 2002 Architectural License for Germany; member of the BDA; member of the German Academy for Urban Development and Environmental Planning. 2003 Founded Werner Sobek New York; peer review engineer in Chicago; exhibition »Beyond Materiality«, Tokyo Exhibition Centre, Tokyo, Japan. 2004 Exhibition »Beyond Materiality – China Tour«, Beijing, Houa Zhong Keij, Beifang Jidaoda, Suzhou, Zhejiang, Fuzhou, Kanton, Shenzhen; exhibition »show me the future – engineering and design by werner sobek« in the Pinakothek der Moderne, Munich, Germany. 2006 Member of the Council on Research and Technology of the University of Graz, Austria; President of the Academic Council of the HafenCity University, Hamburg, Germany. 2007 Visiting professor at Harvard University, Cambridge, Massachusetts; founded WSGreenTechnologies, Werner Sobek Moskwa, and Werner Sobek Cairo. 2008 Founded Werner Sobek Dubai.

1953 geboren in Aalen, Württemberg. 1974–1980 Bauingenieur- und Architekturstudium an der Universität Stuttgart. 1980–1986 Wissenschaftlicher Mitarbeiter am Sonderforschungsbereich »Weitgespannte Flächentragwerke« an der Universität Stuttgart. 1987 Promotion im Bauingenieurwesen an der Universität Stuttgart. 1987–1991 Mitarbeiter im Ingenieurbüro Schlaich, Bergermann & Partner, Stuttgart. 1991–1994 Professor an der Universität Hannover (Nachfolge Bernd Tokarz) und Leiter des Instituts für Tragwerksentwurf und Bauweisenforschung. 1992 Gründung Werner Sobek Stuttgart. 1994 Professor an der Universität Stuttgart (Nachfolge Frei Otto), Direktor des Instituts für Leichte Flächentragwerke. 1998–2006 Mitglied des Vorstands der Ingenieurkammer Baden-Württemberg. 1998 Ernennung zum Prüfingenieur für Baustatik für alle Fachrichtungen. 1999 Gründung Werner Sobek Design; Ausstellung »Archineering«, Städtisches Museum Leverkusen. 2000 Übernahme eines zweiten Lehrstuhls als Nachfolger von Prof. Dr. Jörg Schlaich; Gründung des Instituts für Leichtbau Entwerfen und Konstruieren (ILEK); Shanghai Engineering License. 2000–2001 Gastprofessor an der Architekturfakultät der Universität Graz. 2001 Gründung von Werner Sobek Frankfurt; Structural Engineering License für den US-Bundesstaat Illinois. 2002 Mitglied der Architektenkammer Baden-Württemberg; Mitglied des BDA; Mitglied der Deutschen Akademie für Städtebau und Landesplanung. 2003 Gründung von Werner Sobek New York; Peer Review Engineer in Chicago; Ausstellung »Beyond Materiality«, Messe Tokio. 2004 Ausstellung »Beyond Materiality – China Tour«, Peking, Houa Zhong Keij, Beifang Jidaoda, Suzhou, Zhejiang, Fuzhou, Kanton, Shenzhen; Ausstellung »show me the future – engineering and design by werner sobek« in der Pinakothek der Moderne, München. 2006 Mitglied des Forschungs- und Technologiebeirats der TU Graz; Vorsitzender des Hochschulrats der HafenCity-Uni, Hamburg. 2007 Gastprofessor an der Harvard University, Cambridge, Massachusetts; Gründung von WSGreenTechnologies, Werner Sobek Moskwa und Werner Sobek Cairo. 2008 Gründung von Werner Sobek Dubai.

2008 Design Award of the Federal Republic of Germany
2007 Finalist for the Mies van der Rohe Award
AIA Chicago Distinguished Building Award, Citation of Merit
AIA Chicago Distinguished Building Award, Honor Award
2006 Hugo Häring Award
Architecture + Technology Award, Distinction
Balthasar Neumann Award
Art Directors Club (ADC) Award, Gold
Award of the Deutscher Designer Club (DDC), Silver
German Critics Award Architecture
German Steel Construction Award
Oscar Faber Award
AIA Distinguished Building Award, Citation of Merit (2 ×)
2005 Detail Award
Auguste Perret Prize of the UIA
NYC Art Commission Award for Excellence in Design
»Good Buildings« Award BDA Stuttgart / BDA Heilbronn
Fazlur Rahman Khan Medal
2004 German Steel Construction Award, Distinction
2003 iF design award
Hugo Häring Award
AIA Distinguished Building Award, Citation of Merit
AIA Distinguished Building Award, Honor Award
2002 Innovation Award for Architecture
2001 DuPont Benedictus Award, Exceptional Merit
CBC Merit Award Finalist
2000 German Steel Construction Award, Distinction
Fritz Schumacher Award of the Alfred-Töpfer-Stiftung
»Building of the Year« Award of the AIV Hamburg
German Gluelam Award
European Gluelam Award
1998 DuPont Benedictus Award, Honorable Mention
IFAI Design Award
1997 Timber Construction Award Baden-Württemberg
1991 German Engineering Award, Distinction
1989 Hubert Rüsch Award of the Deutscher Beton-Verein
1983 Fazlur R. Khan Award of the Skidmore, Owings & Merrill Foundation, New York
1981 Award of the Friends of the University of Stuttgart for special scientific achievements

Editors and concept/Herausgeber und Konzeption
Frank Heinlein, Maren Sostmann
Copy/Text Frank Heinlein
Design, layout and typesetting/Gestaltung, Layout und Satz
büro uebele visuelle kommunikation, Beate Gerner, Maja Mory, Silke Sabow, Andreas Uebele
Illustrations/Abbildungen H.G. Esch (9–10, 31, 33, 39, 41, 45–47, 57–59, 61, 99, 109, 127), Andreas Keller (12–13, 42, 73, 133), Rainer Viertlböck (15–17, 21–22, 24, 26–27, 63–65), SOM (19), Frank Rümmele (29, 49–50, 101), Roland Halbe (35–36, 105–107, 122, 143, 145), Petra Böttcher (53), Werner Sobek (55, 69–70, 81–82, 84–85, 115, 129, 135, 141, 151), Florian Holzherr (75), Zooey Braun (77–79, 87–89, 95, 111–113, 139, 147–149), Christian Gahl (91), Uwe Walter (92–93), Johannes Marburg (103), Peter Studer (117), Christoph Gebler (119), Christian Richters (121), Brigida Gonzalez (124–125), Wolfgang Schmidberger (137)
Typeface/Schrift FuturaTBook
Paper/Papier UPM fine SC, 150 g
German proofreading/Deutsches Lektorat Martina Fiess
English proofreading/Englisches Lektorat Beverley Locke
Repro/Repro immedia23
Printing/Druck Leibfarth & Schwarz
Binding/Bindung Verlagsbuchbinderei Karl Dieringer

Distribution/Vertrieb
avedition GmbH
Publishers for architecture and design
Königsallee 57
71638 Ludwigsburg, Germany
www.avedition.com
contact@avedition.com

ISBN 978-3-89986-087-0